Atmospheric Sciences

Atmospheric Sciences

Edited by **Smith Paul**

New York

Published by Callisto Reference,
106 Park Avenue, Suite 200,
New York, NY 10016, USA
www.callistoreference.com

Atmospheric Sciences
Edited by Smith Paul

International Standard Book Number: 978-1-63239-628-0 (Hardback)

Printed in the United States of America.

Contents

Preface

The various sub-fields of atmospheric sciences along with latest technological progress that have future implications are glanced at in this book. Some of the diverse topics covered in this book address the varied branches that fall under this category. It studies earth's atmosphere, processes, the effects of the atmosphere on the other systems, and the effects that other systems have on the atmosphere and includes meteorology, climatology, and aeronomy. It includes some of the vital pieces of work being conducted across the world, on various topics related to atmospheric sciences. It presents researches and studies performed by veterans across the globe. This book aims to equip students and experts with the advanced topics and upcoming concepts in this area.

This book unites the global concepts and researches in an organized manner for a comprehensive understanding of the subject. It is a ripe text for all researchers, students, scientists or anyone else who is interested in acquiring a better knowledge of this dynamic field.

I extend my sincere thanks to the contributors for such eloquent research chapters. Finally, I thank my family for being a source of support and help.

Editor

Seasonal Variations in Health Hazards from Polycyclic Aromatic Hydrocarbons Bound to Submicrometer Particles at Three Characteristic Sites in the Heavily Polluted Polish Region

Barbara Kozielska [1],*, Wioletta Rogula-Kozłowska [2] and Krzysztof Klejnowski [2]

[1] Department of Air Protection, Silesian University of Technology, 2 Akademicka St.,
44-100 Gliwice, Poland

[2] Institute of Environmental Engineering, Polish Academy of Sciences, 34 M. Skłodowskiej-Curie St.,
41-819 Zabrze, Poland; E-Mails: wioletta.rogula-kozlowska@ipis.zabrze.pl (W.R.-K.);
krzysztof.klejnowski@ipis.zabrze.pl (K.K.)

* Author to whom correspondence should be addressed; E-Mail: barbara.kozielska@polsl.pl

Academic Editor: Huiting Mao

Abstract: Suspended particles with aerodynamic diameters not greater than 1 μm (PM$_1$) were sampled at the urban background; regional background; and urban traffic points in southern Poland. In total, 120 samples were collected between 2 August 2009 and 27 December 2010. Sixteen polycyclic aromatic hydrocarbons (PAHs) were determined in each sample. The samples were collected with a high volume sampler (Digitel). Afterwards, they were chemically analyzed with a gas chromatograph equipped with a flame ionization detector (Perkin Elmer Clarus 500). The mean concentration values of the PAH sum (ΣPAH) and particular PAHs; the percentages of carcinogenic PAHs in total PAHs (ΣPAH$_{carc}$/ΣPAH); carcinogenic equivalent (CEQ); mutagenic equivalent (MEQ); and TCDD-toxic equivalent (TEQ) were much higher in the winter (heating) season than in the summer (non-heating) one. For both periods, the resulting average values obtained were significantly higher (a few; and sometimes a several dozen times higher) in the researched Polish region than the values observed in other areas of the world. Such results indicate the importance of health hazards resulting from PM$_1$ and PM$_1$-bound PAHs in this Polish area.

Keywords: PAH; organic matter; PM$_1$; health hazard; mutagenicity; carcinogenicity; TEF; municipal emission

1. Introduction

The degree to which the ambient particulate matter-bound (PM-bound) polycyclic aromatic hydrocarbons (PAHs) are hazardous to health depends on: PAH concentrations both in the ambient air and in the PM mass; their mass distributions in respect to the particle size; and the physicochemical properties of PM.

Despite that fact that almost every PM-bound PAH, and PM itself, strongly affects human health, the ambient concentrations of only a few of them are limited. The following PAHs (16 congeners) are considered as a priority due to their health effects: naphthalene (Na), acenaphthylene (Acy), acenaphthene (Ace), fluorene (Flu), phenanthrene (Ph), anthracene (An), fluoranthene (Fl), pyrene (Py), benzo[a]anthracene (BaA), chrysene (Ch), benzo[b]fluoranthene (BbF), benzo[k]fluoranthene (BkF), benzo[a]pyrene (BaP), indeno[1,2,3-cd]pyrene (IP), dibenzo[ah]anthracene (DBA) and benzo[ghi]perylene (BghiP). According to the International Agency for Research on Cancer (IARC), 48 PAHs can be carcinogenic for humans or animals. The strength of their carcinogenicity grows with their molecular weight. However, it is still unknown if the carcinogenicity of PAHs (which always occur in the mixture form and never as a single compound in the air) may be ascribed to the individual hydrocarbons or if it is rather the concerted effect of some PAHs combined together.

BaP is a well-studied five-ring hydrocarbon. It is particularly important to the environmental toxicology as one of the most mutagenic and carcinogenic hydrocarbons [1–4]. It also constitutes a basis for defining the toxic (carcinogenic) equivalence factor (TEF) and the carcinogenic equivalent (CEQ) for other PAHs. Namely, a TEF for a PAH is defined as relative to the TEF of BaP (assumed to be 1). The CEQ of a PAH group is the linear combination of the TEF and the ambient concentrations of the PAHs from this group. TEF expresses absolute toxicity of a particular PAH, whereas CEQ specifies carcinogenicity of a PAH group in the air [5]. The so-called mutagenic equivalent (MEQ) or the TCDD-toxic equivalent (TEQ) can also be useful as indicators to assess the PAH mixture health effect [6].

In Poland, and particularly in its southern part, PM comes mainly from energy production and municipal and industrial sources [7–10]. To some degree, Southern Poland is representative of those areas in East and Central Europe where power and heat are produced mainly from fossil fuels (coal), and the road traffic is still not so dense as in the western and northern parts of Europe, or even in the USA ten years ago. In other urbanized regions of Western Europe, air pollution with PM is mainly due to road traffic [11]. The PM$_1$-bound PAHs, the concern of this paper, are well enough recognized both in Poland as well as in other European countries.

The following study presents the results of the research into the ambient concentrations of 16 PM$_1$-bound PAHs (Na, Acy, Ace, Flu, Ph, An, Fl, Py, BaA, Ch, BbF, BkF, BaP, IP, DBA and BghiP) at three locations in Silesia (southern Poland). The measurement points were selected due to their specific emission conditions. Seasonal variations of 16 PM$_1$-bound PAHs and their mutual correlations in the seasons were discussed. Additionally, the seasonal variations in the exposure to the mixture of the PM₁-bound PAHs were assessed with selected indicators ($\sum$PAH$_{carc}$/$\sum$PAH, CEQ, MEQ and TEQ).

2. Materials and Methods

Twenty-four-hour samples of PM₁ were collected (with a high volume DIGITEL DHA-80 sampler (flow rate 30 m³·h⁻¹) onto the quartz fiber filters (Whatman™ 1851-150 Grade QM-A, Piscataway, NJ, USA) at three sites in southern Poland (Figure 1); prior to sampling, the filters were baked at 600 °C for at least 6 h to remove any traces of organics.

Figure 1. Sampling sites.

The first point met the requirements for the so-called urban background site (UB); the second one represented the regional background site (RB); and the third one was the so-called urban traffic site (urban site directly affected by road traffic, UT). UB was located in a residential district in the western part of Katowice, approx. 3.2 km west of the city center. There were blocks of flats, commercial areas, and a railway line in its neighborhood. A post-mining terrain could be found at some distance off. UT was located near the A4 highway—Almost on its shoulder, approx. 1.5 km south of the city center. The volume of traffic was about 30,000 vehicles per day at this point. RB was located in Złoty Potok (commune of Janów), approx. 20 km south-east of Częstochowa and 25 km north of Zawiercie. It was surrounded by meadows and agricultural lands. Several wood-heated chalets and a forester's house were approx. 150 m away.

The research took place between 2 August 2009 and 27 December 2010. There were two measurement campaigns at each sampling point held in each heating (January–March and October–December) and non-heating (April–September) season. Six to fourteen (usually 10) 24-h samples of PM_1 were collected in one campaign. Importantly, the campaigns were not consecutive at a given point. For example, in the heating season, the measurements were conducted in the following order: RB, UB and UT. Afterwards, the sampling was conducted in the same way. Altogether, there were 40 24-h samples collected for each point (120 samples in total).

The sampled dust mass was determined gravimetrically (Sartorius balance, resolution 0.01 g; Goettingen, Germany) according to the CSN EN 14907 standard (Ambient air quality—Standard gravimetric measurement method for the determination of the $PM_{2.5}$ mass fraction of suspended particulate matter). Before each weighing, filters were conditioned in a weighing room (min. 48 h; air temperature of 20 ± 1 °C; and air relative humidity of $50\% \pm 5\%$). After weighing, the filters were put into petri dishes wrapped light-tightly in aluminum foil and stored in a freezer at -18 °C till the analysis.

The dust collected on a filter was extracted with 3×10 cm^3 dichloromethane (DCM) for 30 min under ultrasonic agitation. The extract was percolated, washed and dried in helium atmosphere. The dry residue was dissolved in propanol-2 ($CH_3CH(OH)CH_3$), next distilled water was added to receive alcohol/water volume ratio 15/85. For selective purification, the obtained samples were solidified (SPE) via extraction in columns filled with octadecylsilane – C-18 (Supelclear™ ENVI-18 Tubes, Supelco, Bellefonte, PA, USA). Before the extraction, the columns were conditioned with methanol and next with a mixture of propanol-2/water (15/85). Subsequently, the sample was passed through the columns. After the extraction, the bed was dried under vacuum. The fraction of PAHs was eluted with two portions of 0.5 cm^3 of DCM. The extract of PAHs was thickened in helium to the volume of 0.1 cm^3.

The SPE extracts were analyzed on a Clarus 500 Perkin Elmer gas chromatograph (PerkinElmer, Inc., Waltham, MA, USA) equipped with an auto-sampler. The compounds were separated on a capillary column (Restek RTX-5, 5% phenyl methyl siloxane, 30 m $\times$ 0.32 mm $\times$ 0.25 μm). The carrier gas (helium) flow in the column was maintained at the constant rate of 1.5 cm$^3 \cdot$min^{-1}. The 3 μL samples were introduced using splitless injection, the temperature of the injector was 240 °C. A flame ionization detector (FID) was used. For the PAHs analysis, the initial temperature of the oven, 60 °C, was held for 4 min, then the temperature grew at 10 °C$\cdot$min^{-1} to 280 °C and held for 14 min. The flow rates of hydrogen and air in the detector were 45 cm$^3 \cdot$min^{-1} and 450 cm$^3 \cdot$min^{-1}, respectively; the FID's temperature was 280 °C.

The calibration curves for the quantitative analysis were prepared for 16 standard PAHs. The linear correlation between the peak surface areas and PAH concentrations was checked within the range of 1–20 μg$\cdot$mL^{-1} (correlation coefficients: 0.99; PAH Mix PM-611 Ultra Scientific standard at the concentration of 100 μg$\cdot$mL^{-1} for each PAH in DCM).

The analysis of each campaign sample series was accompanied with the blank sample analysis. It consisted in the application of the whole analytical procedure to a clean quartz fiber filter. The blank result was used to adjust the PAH concentration, but only if the blank exceeded 10% of the PAH concentration. The limits of detection (LODs), obtained from the statistical development of the blank results (10–11 blanks for each PAH; PN-EN 15549 standard), were between 0.01 and 0.03 ng·m^{-3} (average air flow rate = 700 m^3/24 h).

The method performance was verified through analyzing the NIST SRM 1649b reference material and comparing the results with the certified concentrations of the investigated PAHs. The standard recoveries were from 92% to 111%.

The cumulative health hazard from the PAH mixture was expressed quantitatively as the carcinogenic equivalent (CEQ) or mutagenic equivalent (MEQ) relative to the carcinogenicity or mutagenicity of BaP, respectively, or as the TCDD-toxic equivalent (TEQ) relative to the 2,3,7,8-tetrachlorodibenzo-p-dioxin toxicity. The indicators were computed for each site and, separately, for the heating and non-heating periods: CEQs as linear combinations of the concentrations of PM$_1$-related PAHs and their toxicity equivalence factors (TEF); MEQs as linear combinations of the concentrations and the PAH minimum mutagenic concentrations (MMC); TEQs as linear combinations of the concentrations of the PM$_1$-related PAHs and their TCDD-TEF. The values of TEF, MMC and TCDD-TEF for particular PAHs were taken from Nisbet and LaGoy [5], Durant et al. [2] and Willett et al. [12]. Additionally, the proportion of the concentration sum of seven carcinogenic PAHs (Ch, BaA, BbF, BkF, BaP, DBA and IP - ΣPAH$_{carc}$ [13]) to the concentration sum of the 16 determined PAHs was calculated. The closer the value of ΣPAH$_{carc}$/ΣPAH was to 1, the more hazardous ΣPAH was to humans.

3. Results and Discussion

In the heating season, the mean PM$_1$ concentrations were 16.37 µg·m^{-3}, 40.70 µg·m^{-3}, 41.55 µg·m^{-3} at RB, UB and UT, respectively. They were 1.6-2.3 times higher than the mean PM$_1$ concentrations in the non-heating season (Table 1). When compared to the European concentrations, the values were high. For example, winter concentrations of PM$_1$ observed in Brno and Šlapanice (Czech Republic) in 2009 were 19.9 µg·m^{-3} and 19.2 µg·m^{-3}, respectively [14]. Perrone et al. [15] measured even lower PM$_1$ concentrations (13 µg·m^{-3}) in southern Italy.

Fine PM concentrations in Asian cities were similar to the Polish ones or even higher. In Tabriz (Iran), the PM$_1$ concentrations measured in the urban and industrial-suburban sites were 28.4 µg·m^{-3} and 31.4 µg·m^{-3}, respectively [16]. In Shanghai (China), the mean PM$_1$ concentration was 78.9 µg·m^{-3} [17]. In Beijing (China), the PM$_{2.5}$ concentration varied and depended on the measurement point location. Nonetheless, it was never lower than 100 µg·m^{-3} [18]. In Agra (India), the mean PM$_{2.5}$ concentration differed at various locations and ranged between 91.2 and 308.3 µg·m^{-3} [19].

Similarly to PM$_1$ concentrations, concentrations of the 16 PAH (ΣPAH) sum and particular PAHs varied and depended on the season. The definitely higher 24-h ΣPAH concentrations were observed in the heating season, which was visible in the mean concentrations ranging between 23.1 ng·m^{-3} (RB) and 186.1 ng·m^{-3} (UT). In the non-heating season, the values ranged between 18.6 ng·m^{-3} (RB) and 56.0 ng·m^{-3} (UT) (Table 1). The biggest differences in the 24-h and mean seasonal concentrations of ΣPAH were found for UB, whereas the lowest values were observed at RB. The heating season ΣPAH/non-heating season ΣPAH ratios were 1.2, 4.6 and 3.3 at RB, UB and UT, respectively.

Table 1. The statistics of 24-h concentrations of PM$_1$ ($\mu g \cdot m^{-3}$) and PM$_1$-bound PAHs (ng·m^{-3}) at three locations in two measurement periods.

| | Regional Background (RB) | | | | | | Urban Background (UB) | | | | | | Traffic Point (UT) | | | | | |
| | Heating Season (N=20) | | | Non-Heating Season (N=20) | | | Heating Season (N=20) | | | Non-Heating Season (N=20) | | | Heating Season (N=20) | | | Non-Heating season (N=20) | | |
	Min / Max	Avrg±SD	50%*	Min / Max	Avrg±SD	50%	Min / Max	Avrg±SD	50%	Min / Max	Avrg±SD	50%	Min / Max	Avrg±SD	50%	Min / Max	Avrg±SD	50%
PM$_1$	3.49 / 71.41	16.37±14.42	14.40	4.82 / 16.06	10.32±3.51	10.17	17.33 / 73.59	40.70±14.82	36.64	7.99 / 34.86	20.83±8.50	24.24	20.50 / 88.34	41.55±16.56	36.60	11.83 / 27.72	18.40±4.48	17.17
16PAH	6.74 / 82.25	23.10±17.82	17.49	3.48 / 45.32	18.57±11.75	13.86	31.32 / 400.14	138.74±87.07	116.28	16.18 / 58.59	30.26±11.15	27.28	20.23 / 695.67	186.12±162.63	132.42	8.23 / 179.89	56.02±43.79	42.31
Na	0 / 0	0±0	0	0 / 0.48	0.02±0.11	0	0 / 0	0±0	0	0 / 0	0±0	0	0 / 0	0±0	0	0 / 0	0±0	0
Acy	0 / 3.40	0.51±0.81	0.11	0 / 1.74	0.35±0.50	0	0 / 0.39	0.02±0.09	0	0 / 4.52	0.62±1.04	0.32	0 / 3.36	0.17±0.75	0	0 / 0.42	0.02±0.09	0
Ace	0 / 5.29	0.35±1.18	0	0 / 1.16	0.12±0.28	0	0 / 1.26	0.28±0.38	0	0 / 16.32	2.64±3.87	0.80	0 / 2.20	0.11±0.49	0	0 / 1.45	0.14±0.38	0
Flu	0.40 / 14.82	2.55±3.28	1.61	0 / 7.97	2.81±2.60	1.67	0.33 / 12.11	2.88±3.37	1.10	0.40 / 1.44	0.82±0.29	0.76	0 / 15.51	3.54±4.16	2.16	0.77 / 25.06	6.11±7.44	2.85
Ph	0 / 2.63	0.64±0.92	0	0 / 2.19	0.93±0.78	1.10	2.53 / 32.67	9.98±8.29	6.64	0.53 / 16.97	3.62±4.37	1.81	0 / 97.13	15.55±23.31	7.54	0 / 18.16	1.69±3.93	0.93
An	0 / 5.42	0.64±1.34	0	0 / 6.61	0.75±1.58	0	0 / 16.28	3.53±4.73	1.73	0.28 / 12.91	2.16±3.28	0.90	0 / 40.88	6.31±9.85	2.63	0.35 / 17.24	4.84±5.45	2.07
Fl	1.29 / 29.88	3.42±6.25	1.96	1.33 / 6.37	2.65±1.53	2.07	3.02 / 71.89	23.36±17.22	18.76	1.23 / 6.01	2.01±1.01	1.79	1.86 / 109.03	26.69±28.30	15.20	0.55 / 8.91	3.62±2.02	3.60
Py	0 / 4.73	1.86±1.39	1.73	0 / 4.24	1.64±1.14	1.49	3.53 / 60.49	20.30±14.40	16.02	0.40 / 6.24	1.87±1.47	1.17	1.67 / 90.12	21.51±22.04	13.89	0.66 / 78.64	10.90±19.18	2.98
BaA	0 / 13.16	2.05±2.98	0.95	0 / 9.54	1.57±2.63	0	3.78 / 51.28	17.67±11.37	14.25	5.08 / 10.73	7.81±1.83	7.75	1.72 / 76.14	19.02±17.22	14.03	2.60 / 24.01	10.41±4.46	9.44
Ch	0.62 / 8.60	3.26±2.15	2.38	0.76 / 5.66	2.81±1.78	2.71	5.19 / 37.56	17.30±7.62	17.30	1.70 / 3.82	2.48±0.50	2.34	2.24 / 59.95	19.85±13.83	18.24	1.01 / 12.68	3.86±2.34	3.23
BbF	0 / 6.61	1.55±1.46	1.25	0 / 3.34	0.80±1.13	0.42	3.17 / 24.64	9.95±4.92	8.89	0 / 1.78	0.76±0.67	0.62	1.29 / 35.88	14.23±8.66	11.86	0 / 6.94	3.38±2.46	3.83

Table 1. *Cont.*

	Regional Background (RB)						Urban Background (UB)						Traffic Point (UT)					
	Heating Season (N = 20)			Non-Heating Season (N = 20)			Heating Season (N = 20)			Non-Heating Season (N = 20)			Heating Season (N = 20)			Non-Heating season (N = 20)		
	Avrg ± SD	50% *	Min / Max	Avrg ± SD	50%	Min / Max	Avrg ± SD	50%	Min / Max	Avrg ± SD	50%	Min / Max	Avrg ± SD	50%	Min / Max	Avrg ± SD	50%	Min / Max
BkF	1.24 ± 0.92	1.18	0 / 3.09	0.68 ± 0.96	0.22	0 / 2.98	10.74 ± 5.21	9.51	3.22 / 26.09	1.87 ± 1.58	1.56	0 / 7.13	13.80 ± 7.77	11.97	1.79 / 34.51	0.97 ± 1.38	0.68	0 / 6.01
BaP	4.03 ± 5.97	2.40	0.80 / 28.23	2.46 ± 2.61	1.51	0 / 9.52	12.48 ± 6.36	11.29	4.52 / 32.28	2.97 ± 1.39	3.15	0.41 / 5.14	14.27 ± 11.27	11.87	1.48 / 46.70	4.73 ± 3.92	4.13	0 / 19.45
IP	0.49 ± 1.25	0	0 / 5.52	0.38 ± 1.17	0	0 / 5.15	5.03 ± 4.18	3.95	0.77 / 18.70	0.35 ± 0.36	0.43	0 / 1.07	5.57 ± 3.83	5.89	0 / 13.36	0.51 ± 1.48	0	0 / 6.59
DBA	0.21 ± 0.58	0	0 / 2.37	0.37 ± 1.19	0	0 / 5.25	0.27 ± 0.85	0	0 / 3.71	0.24 ± 0.38	0	0 / 0.94	17.21 ± 12.92	11.59	0 / 47.91	4.77 ± 12.80	0.40	0 / 57.64
BghiP	0.30 ± 0.59	0	0 / 1.82	0.24 ± 1.06	0	0 / 4.73	4.95 ± 4.43	3.98	0 / 17.17	0.05 ± 0.16	0	0 / 0.59	8.29 ± 5.82	7.97	0 / 23.00	0.09 ± 0.28	0	0 / 1.12

* median.

The mean concentrations of nearly all the compounds differed in each season at each measurement point. For example, for UT, the mean concentrations of 16 PAHs in the heating season were in the decreasing order Fl > Py > Ch > BaA > DBA > Ph > BaP > BbF > BkF > BghiP > An > IP > Flu > Acy > Ace (0.1–26.7 ng·m^{-3}); whereas, the order was Py > BaA > Flu > An > DBA > BaP > Ch > Fl > BbF > Ph > BkF > IP > Ace > BghiP > Acy (0.02–10.9 ng·m^{-3}) in the non-heating season.

The profile of 16 PM$_1$-bound PAHs is illustrated in Figures 2–4. In RB, Ch, Fl, BaP, Flu, Py had the biggest total percentages in ∑PAH (67.9% and 72.8% of the ∑PAH mass in the heating and non-heating seasons, respectively). In the heating season in UB, BaA, Fl, Ch, BaP and Py constituted 66.1% of the ∑PAH mass. BaA, BaP, Ph and Ch were dominant in the non-heating season (total value = 58.0%). The PAH profile in UT was slightly different - BaA, Py, Fl, Ch, DBA, BaP and BbF made 74.4% of the ∑PAH mass. In the heating season, Fl, DBA, Ch, Py, BaA, BkF, BbF and BaP were responsible for 82.6% of the ∑PAH mass. In the non-heating season, BaA, Py, BaP, Flu and An constituted 65.9% of the ∑PAH mass (BaA made 22.8%).

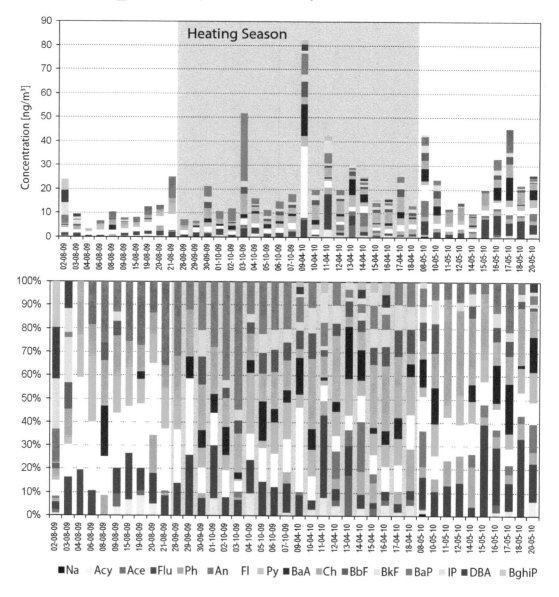

Figure 2. Twenty-four-hour concentrations of 16 PAHs contained in PM$_1$ and percentages of specific PAHs in the PAH sum at the regional background point (RB).

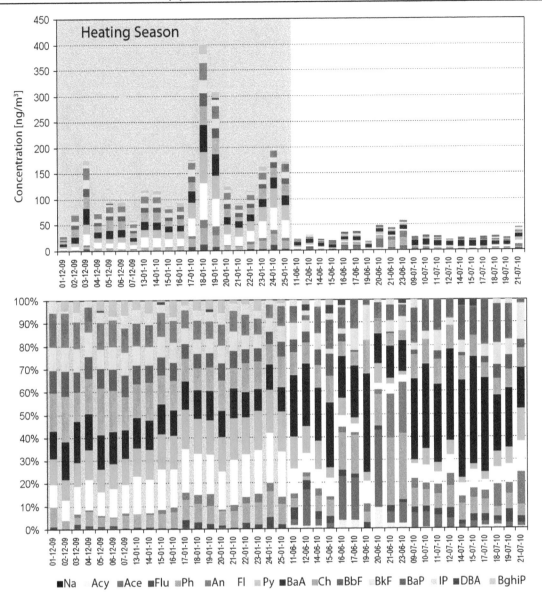

Figure 3. Twenty-four-hour concentrations of 16 PAHs contained in PM₁ and percentages of specific PAHs in the PAH sum at the urban background point (UB).

On several winter days (03.10.2009 and 9.04.2010 at RB, 17–19.01.2010 at UB, 29.10.2010 at UT), some PAHs had unusually high mass contributions to ∑PAH and ambient concentrations and caused very high ∑PAH ambient concentrations. At each point they were BaA, Py, Fl, Ch, BaP, and BbF, markers of the emissions from coal combustion and incineration [4]. Such high ambient concentrations of these compounds in the high pollution episode days may be accounted for by intensified fuel combustion (heating season) and also by the meteorological conditions favoring accumulation of the pollutants in the near-ground layer of the atmosphere.

The differences in the PAH profiles, both between the points and within the surroundings of each measurement point between the seasons, reflected the differences in the PAH origin at the points and in seasons. The finding was corroborated by the differences in the linear correlation coefficients calculated between the concentrations of specific PAHs, the PAH sum and PM₁ for the points and seasons (Figure 5).

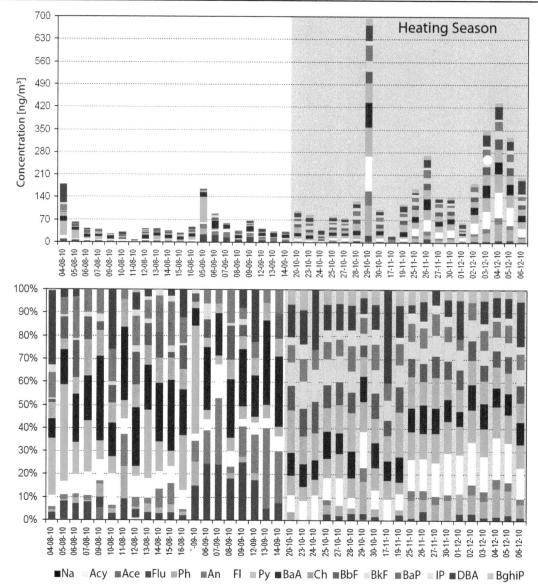

Figure 4. Twenty-four-hour concentrations of 16 PAHs contained in PM$_1$ and percentages of specific PAHs in the PAH sum at the traffic point (UT).

In the non-heating season in UT, RB and UB, there were high values of correlations between the concentrations of heavier PAHs (*i.e.*, BghiP or DBA), which confirmed the dominant influence of the PAH emission from traffic sources [10,18,20,21]. It seems that due to the lack or to the significant reduction of emissions from other sources, traffic and industry (including energy production) were dominant sources of PAHs in the air of southern Poland. In the heating period (particularly in UB and UT), concentrations of most examined compounds were highly correlated. It is possible that the activity of numerous PAH sources was responsible for such a situation. In UB, these were mainly low emissions (coal and biomass combustion in home furnaces) and more intensive (than in summer) energy production (mainly based on the hard and brown coal combustion). In UT, the same sources were active; additionally, there was a strong influence of the traffic emission. In RB, mainly local and dispersed sources and the inflow of pollutants from other, more polluted regions, affected the air pollution with PAHs in both seasons [10,22,23].

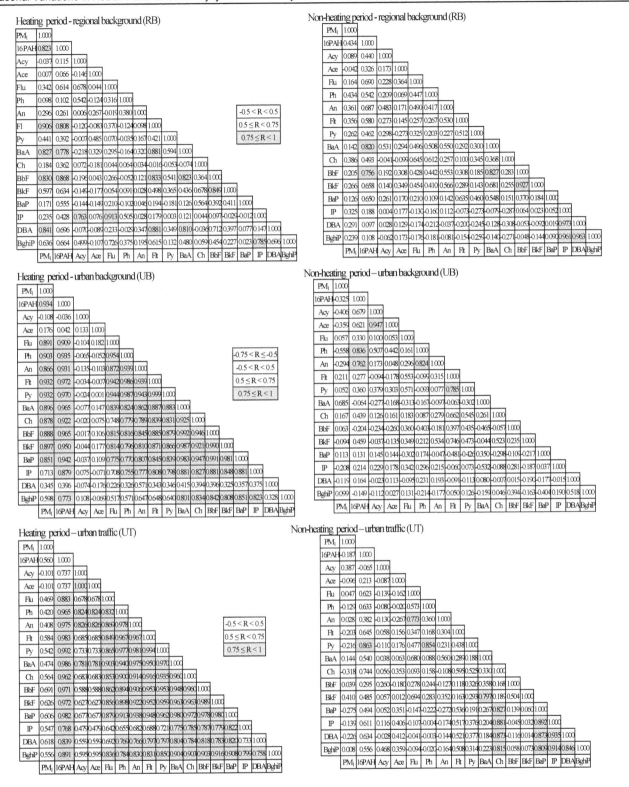

Figure 5. Correlation matrices for 24-h concentrations of PM1, sum of 16 PAHs and particular PAHs in two averaging periods for three locations in Silesia (Poland).

It is worth taking into account the high BaP percentage in the concentration of 16 PAHs. For the examined areas and over the whole measurement period, it was 9%–13%. The ambient concentration of the PM_{10}-bound BaP (BaP bound to particles not greater than 10 μm) has the limit that should not be exceeded; the yearly permissible ambient concentration of PM_{10}-bound BaP is 1 ng·m^{-3} [24,25].

Consequently, the mean PM_1-bound BaP concentrations in RB, UB and UT were very high. The values were 3.24 ng·m^{-3} (4.03 ng·m^{-3}—Heating season; 2.46 ng·m^{-3}—Non-heating season), 7.72 ng·m^{-3} (12.48 ng·m^{-3}—Heating season; 2.97 ng·m^{-3}—Non-heating season) and 9.5 ng·m^{-3} (14.27 ng·m^{-3}—Heating season; 4.73 ng·m^{-3}—Non-heating season) in RB, UB and UT, respectively. Such a finding shows that even though RB was located far away from the anthropogenic emission sources, the BaP and PM_1-bound BaP concentrations exceeded the permissible level more than three times there. In UB in the densely developed and populated area, the permissible BaP value was exceeded nearly eight times, whereas the value was almost 10 times higher in UT. High BaP concentrations were observed both in the heating season and the remaining part of the year. Such a situation was not observed in other European regions even though the permissible mean yearly values were also exceeded in the Czech cities. Nonetheless, the PM_1-bound BaP concentrations did not exceed 0.39 ng·m^{-3} [14]. The research conducted at the roadside and urban background in Madrid (Spain) showed that the BaP concentration in winter was 0.24 ng·m^{-3} and 0.054 ng·m^{-3}, respectively. In summer, the BaP concentrations observed in Madrid (Spain) were even lower [26]. On the other hand, the BaP concentrations as high as those observed in RB, UB and UT are typical for cities in southern Poland and are observed every year. For example, in the urban background in Zabrze (a city 15 km away from Katowice), the mean concentration of the PM_1-bound BaP was 16 ng·m^{-3} in winter 2007 [27].

Table 2 presents BaP concentrations and the following indicators of the exposure to the PAH mixture: carcinogenic equivalent (CEQ), mutagenic equivalent (MEQ), TCDD-toxic equivalent (TEQ), and percentages of seven PAHs determined in this study, whose carcinogenic influence was confirmed in the sum of the determined PAHs ($\sum PAH_{carc}/\sum PAH$). The way in which each indicator was calculated is given under the table. The values in the table were calculated with the results obtained in this study and additionally, results from other world regions taken from the publications quoted in Table 2 [6,10,14,26,28–33].

In the heating season, the mean CEQs for the whole monitoring period were 5.85 ng·m^{-3} (range: 0.01–30.33 ng·m^{-3})—RB; 18.46 ng·m^{-3} (range: 5.78–51.93 ng·m^{-3})—UB; and 106 ng·m^{-3} (range: 19.36–303.78 ng·m^{-3})—UT. They were unquestionably higher than the non-heating season values. It was particularly visible for UT, where CEQ was approx. seven times higher in the heating season than its mean value in the non-heating season.

In the European cities, the CEQ values were lower at the urban and background locations than in Złoty Potok or Katowice (Poland). The values obtained for Asian cities (e.g., Urumqi in China, and Chennai City and Delhi in India) were as high as or higher than those observed in Poland. The values observed for Urumqi and Chennai City in the urban area in winter were 43.18 ng·m^{-3} and 20.72 ng·m^{-3}, respectively. The value for Chennai City was even higher in summer (44.29 ng·m^{-3}) with the BaP concentration of 25.6 ng·m^{-3}. On the other hand, the CEQ value did not exceed 1 ng·m^{-3} in Madrid (Spain), Atlanta (USA), or Bumardas and Chréa (Algeria).

Table 2. Concentration values of BaP, CEQ, MEQ, TEQ and $\sum PAH_{carc}/\sum PAH$ for PM-bound PAHs in various world regions.

City, Country	Sampling Point	Sampling Period	BaP	CEQ	MEQ	TEQ	$\sum PAH_{carc}/\sum PAH$	Reference
				ng·m⁻³		pg·m⁻³		
Złoty Potok, Poland [1]	regional background	heating season 2009/2010	4.03	5.85	4.87	7.36	0.52	This study
Katowice, Poland [1]	urban background		12.48	18.46	20.47	91.95	0.56	
	traffic point		14.27	106.00	29.54	147.79	0.63	
Złoty Potok, Poland [1]	regional background	non-heating season 2009/2010	2.46	4.50	4.29	13.80	0.50	
Katowice, Poland [1]	urban background		2.97	5.29	4.23	13.71	0.59	
	traffic point		4.73	15.46	6.45	20.42	0.55	
Zabrze, Poland [1]	urban background	Winter 2007/2008	16.09	25.04	25.24	116.03	0.66	[6]
Zabrze, Poland [2]	crossroads	June–August 2006	1.1	1.48	1.90	10.29	0.48	[10]
Ruda Śląska [2]	roadside		0.3	1.57	0.87	6.62	0.84	
Brno, Czech Republic [1]	large city	Winter 2009	1.39	5.35	3.19	9.65	0.55	
		Winter 2010	2.82	8.71	5.85	15.81	0.57	
		Summer 2009	0.35	3.17	0.84	2.74	0.58	
		Summer 2010	0.15	0.45	0.29	0.83	0.54	[14]
Šlapanice, Czech Republic [1]	small town	Winter 2009	1.65	5.94	3.72	10.63	0.67	
		Winter 2010	2.84	9.24	5.98	16.56	0.58	
		Summer 2009	0.39	3.08	0.93	2.93	0.58	
		Summer 2010	0.15	0.35	0.27	0.79	0.57	
Madrid, Spain [1]	roadside	Winter 2009	0.24	0.51	0.51	2.55	0.47	
	urban background		0.054	0.12	0.11	0.62	0.56	[26]
	roadside	Summer 2009	0.034	0.12	0.09	0.43	0.47	
	urban background		0.022	0.10	0.05	0.26	0.57	

Table 2. *Cont.*

City, Country	Sampling Point	Sampling Period	BaP	CEQ ng·m⁻³	MEQ	TEQ pg·m⁻³	∑PAH$_{carc}$/∑PAH	Reference
Bumardas, Algeria [1]	urban		0.111	0.43	0.26	0.37	0.44	
Rouiba-Réghaia, Algeria [1]	industrial district	October 2006	0.296	1.48	0.70	1.33	0.62	[28]
Chréa, Algeria [1]	forested mountains		0.018	0.11	0.04	0.08	0.39	
Florence, Italy [2]	urban background	Cold 2009–2010	0.47	2.72	0.99	5.39	0.78	
	urban traffic		1.0	5.43	2.17	11.79	0.74	[29]
Livorno, Italy [2]	suburban background		0.20	0.83	0.44	2.10	0.63	
Florence, Italy [2]	urban background	Warm 2009–2010	0.049	0.41	0.14	1.01	0.61	
	urban traffic		0.21	1.54	0.54	3.47	0.60	
Livorno, Italy [2]	suburban background		0.02	0.09	0.07	0.37	0.46	
Urumqi, China [2]	urban	Winter 2010/2011	2.21	43.18	13.39	63.40	0.77	[30]
		Autumn 2010	0.53	10.09	2.56	12.77	0.93	
Chennai City, India [2]	commercial region	Winter 2009/2010	6.5	13.44	25.77	111.22	0.31	
	urban region		8.1	20.72	37.42	136.13	0.28	[31]
	residential region		16.2	19.68	24.28	37.32	0.13	
	industrial region		24.4	39.78	58.52	242.65	0.21	
Chennai City, India [2]	commercial region	Summer 2009	10.3	16.88	25.96	79.58	0.29	
	urban region		25.6	44.29	77.37	246.68	0.47	
	residential region		6.2	13.85	26.64	94.19	0.39	
	industrial region		8.5	38.37	91.24	516.48	0.33	

Table 2. *Cont.*

City, Country	Sampling Point	Sampling Period	BaP	CEQ ng·m⁻³	MEQ	TEQ pg·m⁻³	∑PAH$_{carc}$/∑PAH	Reference
Delhi, India [2]	traffic point	Winter 2007	9.9	28.92	24.27	125.43	0.67	[32]
		Summer 2007/2008	5.1	17.64	10.60	57.76	0.62	
	urban	April–June 2004	0.04	0.10	0.12	0.48	0.47	[33]
	suburban–highway		0.066	0.15	0.18	0.66	0.51	
	rural		0.015	0.05	0.05	0.23	0.44	
Atlanta, USA [2]	urban	October–December 2004	0.265	0.53	0.59	2.27	0.64	
	suburban–highway		0.492	0.90	0.97	3.52	0.68	
	rural		0.199	0.48	0.46	2.12	0.70	

[1] PM$_1$, [2] PM$_{2.5}$; CEQ = 0.001 × ([Na] + [Acy] + [Ace] + [Flu] + [Ph] + [Fl] + [Py]) + 0.01 × ([An] + [Ch] + [BghiP]) + 0.1 × ([BaA] + [BbF] + [BkF] + [IP]) + 1 × [BaP] + 5 × [DBA]; values of 0.001, 0.01, 0.1 and 1 are the so-called toxic equivalence factors (TEF) for specific PAHs, taken from the study [5]; MEQ = 0.00056 × [Acy] + 0.082 × [BaA] + 0.017 × [Ch] + 0.25 × [BbF] + 0.11 × [BkF] + 1 × [BaP] + 0.31 × [IP] + 0.29 × [DBA] + 0.19 × [BghiP]; values of 0.00056, 0.082, 0.017, 0.25, 0.11, 1, 0.31, 0.29, 0.19 and 0.01 are the so-called minimum mutagenic concentrations (MMC) for specific PAHs, taken from the study [2]; TEQ = 0.000025 × [BaA] + 0.00020 × [Ch] + 0.000354 × [BaP] + 0.00110 × [IP] + 0.00203 × [DBA] + 0.00253 × [BbF] + 0.00487 × [BkF]; values of 0.000025, 0.00020, 0.000354, 0.00110, 0.00203, 0.00253 and 0.00487 are the so-called TCDD-TEF, *i.e.*, toxic equivalency factor relative to 2,3,7,8-tetrachlorodibenzo-p-dioxin TCDD (for TCDD, TEF = 1.0) for specific PAHs, taken from the study [12]; ∑PAH$_{carc}$/∑PAH = ([BaA] + [BaP] + [BbF] + [BkF] + [Ch] + [DBA] + [IP])/([Acy] + [Ace] + [Flu] + [Ph] + [An] + [Fl] + [Py] + [BaA] + [Ch] + [BbF] + [BkF] + [BaP] + [DBA] + [BghiP] + [IP]).

In winter, the MEQ values in UB and UT were also high, *i.e.*, 20.27 n·m^{-3} (range: 6.61–55.21 ng·m^{-3}) and 29.54 ng·m^{-3} (range: 4.55–89.13 ng·m^{-3}), respectively. Similar values were obtained in Zabrze (Poland), Chennai City (India) and Delhi (India). The calculated values were much lower in other locations, particularly in the European ones. For Polish locations, the MEQ values were always higher in the heating season. On the other hand, the MEQ values calculated in summer for Chennai City (India) in the urban and industrial regions were much higher than in winter, *i.e.* 77.37 ng·m^{-3} and 91.24 ng·m^{-3}, respectively.

In the heating season, the CEQ and MEQ at UB were three to four times higher than the values observed at RB. The TEQ values were 12.5 times higher (UB—91.9 pg·m^{-3}; RB—7.4 pg·m^{-3}). At UT (heating season), TEQ was also high (147.8 pg·m^{-3}). A similar situation was observed in Chennai City (India), where TEQ values exceeded 200 pg·m^{-3} and reached 516.5 pg·m^{-3} in the industrial region. In the non-heating season in UB and UT, the mean TEQ values were definitely lower than in the heating season (RB was an exception). In general, the values were still higher than in the European locations.

The mean concentrations of seven carcinogenic PAHs [13] in the heating season were 4.2–4.5 times higher in UT and UB than in the non- heating season. For $\sum PAH_{carc}/\sum PAH$, no significant visible seasonal change was observed at three Polish locations. In the heating season, only in UT was the indicator value slightly higher than in the non-heating season. For the remaining measurement points, the values were similar and ranged between 0.52 and 0.59. For most measurement points for which CEQ, MEQ and TEQ were calculated, the $\sum PAH_{carc}/\sum PAH$ ratio was 0.45–0.60 (Table 2). In the extreme cases, the observed values were 0.84 (roadside; Ruda Śląska, Poland), 0.77 and 0.93 (urban area; Urumqi, China), and 0.78 (urban background; Florence, Italy).

Extremely low values were seen in the research conducted by [31]. The $\sum PAH_{carc}/\sum PAH$ ratios differed from the remaining ones and were 0.13–0.47. The remaining indicators of the exposure to the PAH sum (CEQ, MEQ, TEQ and BaP concentration) were high in comparison to other locations.

4. Conclusions

The detailed analyses helped to reach the following conclusions:

1   over the whole measurement period, mean concentrations of the PM$_1$-bound PAH sum and particular compounds within this group were high at each of the three selected points in Silesia; particularly high values were observed for the heating season;

2   concentrations of the sum of 16 PM$_1$-related PAHs and BaP in the air in Silesia were higher than most values observed in other regions of the world; nonetheless, they did not differ from the concentrations measured in this area previously;

3   the highest concentrations of most PAHs were observed at UT point, both in the heating and in the non-heating seasons;

4   in the typical urban background area, mainly municipal emissions (burning coal, biomass, waste and rubbish in home furnaces) and energy production (mainly based on hard and brown coal combustion) influence PAH concentrations in the air; in the urban site located near highway, the same sources were active as in the urban background area; additionally, there was a strong influence of the traffic emission. In RB, mainly local and dispersed sources and

the inflow of pollutants from other, more polluted regions, affected the air pollution with PAHs in both seasons;

5   high percentage of BaP in the PAH sum (9%–13%) and very high ambient concentrations of the PM$_1$-bound BaP, particularly in the heating season (4–14 ng·m^{-3}), may pose a serious threat to the Silesia inhabitants; the risk does not only concern the residents of large cities and regions located close to important traffic emission sources, it also involves people who dwell in the "clean" areas far away from large urban agglomerations (regional/rural background);

6   in the heating season, the mean CEQs for the whole monitoring period were 5.85 ng·m^{-3} in RB, 18.46 ng·m^{-3} in UB, and 106 ng·m^{-3} in UT; they were unquestionably higher than the non-heating season values and definitely higher than the values obtained in other European regions;

7   MEQ, TEQ and $\sum$PAH$_{carc}$/$\sum$PAH, proposed by the authors as other indicators of the exposure to the PAH mixture, were very high in the Upper Silesian urban area when compared to other regions; the highest indicator values were observed in the heating season;

8   it may be suspected that in Central, Central-East and East Europe traffic is not the primary PAH source; the PM-bound PAHs come mainly from the fossil fuel combustion for heat and power production.

Acknowledgments

The work was partially supported by the Polish Ministry of Science and Higher Education (Grant No. N N523421037).

Author Contributions

The study was completed with cooperation between all authors.

Conflicts of Interest

The authors declare no conflict of interest.

References

1.   Nikolaou, K.; Masclet, P.; Mouvier, G. Sources and chemical reactivity of polynuclear aromatic hydrocarbons in the atmosphere—A critical review. *Sci. Total Environ.* **1984**, *32*, 103–132.

2.   Durant, J.L.; Busby, W.F. Jr.; Lafleur, A.L.; Penman, B.W.; Crespi, C.L. Human cell mutagenicity of oxygenated, nitrated and unsubstituted polycyclic aromatic hydrocarbons associated with urban aerosols. *Mutat. Res. Genetic Toxicol.* **1996**, *371*, 123–157.

3.   Saunders, C.R.; Ramesh, A.; Shockley, D.C. Modulation of neurotoxic behavior in F-344 rats by temporal disposition of benzo(a)pyrene. *Toxicol. Lett.* **2002**, *129*, 33–45.

4.   Ravindra, K.; Sokhi, R.; van Grieken, R. Atmospheric polycyclic aromatic hydrocarbons: Source attribution, emission factors and regulation. *Atmos. Environ.* **2008**, *42*, 2895–2921.

5. Nisbet, I.C.T.; LaGoy, P.K. Toxic Equivalency Factors (TEFs) for Polycyclic Aromatic Hydrocarbons (PAHs). *Regul. Toxicol. Pharmacol.* **1992**, *16*, 290–300.

6. Rogula-Kozłowska, W.; Kozielska, B.; Klejnowski, K. Concentration, origin and health hazard from fine particle-bound PAH at three characteristic sites in Southern Poland. *Bull. Environ. Contam. Toxicol.* **2013**, *91*, 349–355.

7. Majewski, G.; Kleniewska, M.; Brandyk, A. Seasonal variation of particulate matter mass concentration and content of metals. *Pol. J. Environ. Stud.* **2011**, *20*, 417–427.

8. Rogula-Kozłowska, W.; Klejnowski, K.; Rogula-Kopiec, P.; Mathews, B.; Szopa, S. A study on the seasonal mass closure of ambient fine and coarse dusts in Zabrze, Poland. *Bull. Environ. Contam. Toxicol.* **2012**, *88*, 722–729.

9. Kozielska, B.; Rogula-Kozłowska, W.; Pastuszka, J.S. Traffic emission effects on ambient air pollution by $PM_{2.5}$-related PAH in Upper Silesia, Poland. *Int. J. Environ. Pollut.* **2013**, *53*, 245–264.

10. Rogula-Kozłowska, W.; Klejnowski, K.; Rogula-Kopiec, P.; Ośródka, L.; Krajny, E.; Błaszczak, B.; Mathews, B. Spatial and seasonal variability of the mass concentration and chemical composition of $PM_{2.5}$ in Poland. *Air Qual. Atmos. Health* **2014**, *7*, 41–58.

11. Viana, M.; Kuhlbusch, T.A.J.; Querol, X.; Alastuey, A.; Harrison, R.M.; Hopke, P.K.; Winiwarter, W.; Vallius, M.; Szidat, S.; Prévôt, A.S.H.; *et al.* Source apportionment of particulate matter in Europe: A review of methods and results. *J. Aerosol. Sci.* **2008**, *39*, 827–849.

12. Willett, K.L.; Gardinali, P.R.; Sericano, J.L.; Wade, T.L.; Safe, S.H. Characterization of the H4IIE rat hepatoma cell bioassay for evaluation of environmental samples containing Polynuclear Aromatic Hydrocarbons (PAHs). *Arch. Environ. Contam. Toxicol.* **1997**, *32*, 442–448.

13. USEPA (Environmental Protection Agency). *Polycyclic Aromatic Hydrocarbons (PAHs)—EPA Fact Sheet*; National Center for Environmental Assessment, Office of Research and Development: Washington, DC, USA, 2008.

14. Křůmal, K.; Mikuška, P.; Večeřa, Z. Polycyclic aromatic hydrocarbons and hopanes in PM_1 aerosols in urban areas. *Atmos. Environ.* **2013**, *67*, 27–37.

15. Perrone, M.R.; Dinoi, A.; Becagli, S.; Udisti, R. Chemical composition of PM_1 and $PM_{2.5}$ at a suburban site in southern Italy. *Int. J. Environ. Anal. Chem.* **2014**, *94*, 127–150.

16. Gholampour, A.; Nabizadeh, R.; Yunesian, M.; Naseri, S.; Taghipour, H.; Rastkari, N.; Nazmara, S.; Mahvi, A.H. Physicochemical characterization of ambient air particulate matter in Tabriz, Iran. *Bull. Environ. Contam. Toxicol.* **2014**, *92*, 738–744.

17. Shi, Y.; Chen, J.; Hu, D.; Wang, L.; Yang, X.; Wang, X. Airborne submicron particulate (PM_1) pollution in Shanghai, China: Chemical variability, formation/dissociation of associated semi-volatile components and the impacts on visibility. *Sci. Total Environ.* **2014**, *473–474*, 199–206.

18. Duan, J.; Tan, J.; Wang, S.; Chai, F.; He, K.; Hao, J. Roadside, urban, and rural comparison of size distribution characteristics of PAHs and carbonaceous components of Beijing, China. *J. Atmos. Chem.* **2012**, *69*, 337–349.

19. Pachauri, T.; Satsangi, A.; Singla, V.; Lakhani, A.; Maharaj Kumari, K. Characteristics and sources of carbonaceous aerosols in $PM_{2.5}$ during wintertime in Agra, India. *Aerosol Air Qual. Res.* **2013**, *13*, 977–991.

20. Hassan, S.K.; Khoder, M.I. Gas-particle concentration, distribution, and health risk assessment of polycyclic aromatic hydrocarbons at a traffic area of Giza, Egypt. *Environ. Monit. Assess.* **2012**, *184*, 3593–3612.

21. Šišović, A.; Pehnec, G.; Jakovljević, I.; Šilović Hujić, M.; Vacrossed D Signić, V.; Bešlić, I. Polycyclic aromatic hydrocarbons at different crossroads in Zagreb, Croatia. *Bull. Environ. Contam. Toxicol.* **2012**, *88*, 438–442.

22. Tang, N.; Hattori, T.; Taga, R.; Igarashi, K.; Yang, X.; Tamura, K.; Kakimoto, H.; Mishukov, V.F.; Toriba, A.; Kizu, R.; *et al.* Polycyclic aromatic hydrocarbons and nitropolycyclic aromatic hydrocarbons in urban air particulates and their relationship to emission sources in the Pan-Japan Sea countries. *Atmos. Environ.* **2005**, *39*, 5817–5826.

23. Kuo, C.-Y.; Chien, P.-S.; Kuo, W.-C.; Wei, C.-T.; Rau, J.-Y. Comparison of polycyclic aromatic hydrocarbon emissions on gasoline- and diesel-dominated routes. *Environ. Monit. Assess.* **2013**, *185*, 5749–5761.

24. WHO. *Air Quality Guidelines for Europe*; World Health Organization Regional Office for Europe: Copenhagen, Denmark, 2000.

25. EC. Council Directive 2004/107/EC Relating to Arsenic, Cadmium, Mercury, Nickel and Polycyclic Aromatic Hydrocarbons in Ambient Air. Available online: http://eur-lex.europa.eu/LexUriServ/LexUriServ.do?uri=OJ:L:2005:023:0003:0016:EN:PDF (accessed on 17 December 2014).

26. Mirante, F.; Alves, C.; Pio, C.; Pindado, O.; Perez, R.; Revuelta, M.A.; Artiñano, B. Organic composition of size segregated atmospheric particulate matter, during summer and winter sampling campaigns at representative sites in Madrid, Spain. *Atmos. Res.* **2013**, *132–133*, 345–361.

27. Rogula-Kozłowska W.; Kozielska B.; Błaszczak B.; Klejnowski K. The mass distribution of particle-bound PAH among aerosol fractions: A case-study of an urban area in Poland. In *Organic Pollutants Ten Years after the Stockholm Convention—Environmental and Analytical Update*; Puzyn, T., Mostrag-Szlichtyng, A., Eds.; InTech: Rijeka, Croatia, 2012; pp. 163–190.

28. Ladji, R.; Yassa, N.; Balducci, C.; Cecinato, A.; Meklati, B.Y. Distribution of the solvent-extractable organic compounds in fine (PM_1) and coarse (PM_{1-10}) particles in urban, industrial and forest atmospheres on Northern Algeria. *Sci. Total Environ.* **2009**, *409*, 415–424.

29. Martellini, T.; Giannoni, M.; Lepri, L.; Katsoyiannis, A.; Cincinelli, A. One year intensive $PM_{2.5}$ bound polycyclic aromatic hydrocarbons monitoring in the area of Tuscany, Italy. Concentrations, source understanding and implications. *Environ. Pollut.* **2012**, *164*, 252–258.

30. Da Limu, Y.L.M.A.B.; LiFu, D.L.N.T.; Miti, A.B.L.Y.; Wang, X.; Ding, X. Autumn and wintertime polycyclic aromatic hydrocarbons in $PM_{2.5}$ and $PM_{2.5-10}$ from Urumqi, China. *Aerosol Air Qual. Res.* **2013**, *13*, 407–414.

31. Mohanraj, R.; Solaraj, G.; Dhanakumar, S. Fine particulate phase PAHs in ambient atmosphere of Chennai metropolitan city, India. *Environ. Sci. Pollut. Res.* **2011**, *18*, 764–771.

32. Singh, D.P.; Gadi, R.; Mandal, T.K. Characterization of particulate-bound polycyclic aromatic hydrocarbons and trace metals composition of urban air in Delhi, India. *Atmos. Environ.* **2011**, *45*, 7653–7663.

33. Li, Z.; Porter, E.N.; Sjödin, A.; Needham, L.L.; Lee, S.; Russell, A.G.; Mulholland, J.A. Characterization of PM$_{2.5}$-bound polycyclic aromatic hydrocarbons in Atlanta-Seasonal variations at urban, suburban, and rural ambient air monitoring sites. *Atmos. Environ.* **2009**, *43*, 4187–4193.

2

Vertical and Horizontal Polarization Observations of Slowly Varying Solar Emissions from Operational Swiss Weather Radars

Marco Gabella *, Maurizio Sartori, Marco Boscacci and Urs Germann

MeteoSwiss, via ai Monti 146, Locarno Monti CH-6605, Switzerland;
E-Mails: maurizio.sartori@meteoswiss.ch (M.S.); marco.boscacci@meteosvizzera.ch (M.B.);
urs.germann@meteoswiss.ch (U.G.)

* Author to whom correspondence should be addressed; E-Mail: marco.gabella@meteoswiss.ch

Academic Editor: Richard Müller

Abstract: The electromagnetic power that arrives from the Sun in the C-band has been used to check the quality of the polarimetric, Doppler weather radar network that has recently been installed in Switzerland. The operational monitoring of this network is based on the analysis of Sun signals in the polar volume data produced during the MeteoSwiss scan program. It relies on a method that has been developed to: (1) determine electromagnetic antenna pointing; (2) monitor receiver stability; and (3) assess the differential reflectivity offset. Most of the results from such a method had been derived using data acquired in 2008, which was a period of quiet solar flux activity. Here, it has been applied, in simplified form, to the currently active Sun period. This note describes the results that have been obtained recently thanks to an inter-comparison of three polarimetric operational radars and the Sun's reference signal observed in Canada in the S-band by the Dominion Radio Astrophysical Observatory (DRAO). The focus is on relative calibration: horizontal and vertical polarization are evaluated *versus* the DRAO reference and mutually compared. All six radar receivers (three systems, two polarizations) are able to capture and describe the monthly variability of the microwave signal emitted by the Sun. It can be concluded that even this simplified form of the method has the potential to routinely monitor dual-polarization weather radar networks during periods of intense Sun activity.

Keywords: weather radar receivers; monitoring; relative calibration; solar emission

1. Introduction

The idea of using the solar signal detected by ground-based radars during the operational weather scan program has been presented in a series of papers: Huuskonen and Holleman [1] presented the use of Sun signals to determine antenna pointing, which is the residual offset in the angle of elevation and azimuth. Holleman *et al.* [2] extended the method towards quantitative electromagnetic power measurements, in order to monitor radar receiving chain stability. In the case of dual-polarization radars, which are able to measure both horizontal (H) and vertical (V) polarization, the same technique can be used to monitor the presence (and stability) of a residual differential reflectivity offset, as shown, for instance, by Holleman *et al.* [3], who investigated the stability of differential radar reflectivity over three months during periods of quiet Sun emissions. Here, we show recently acquired results from three operational weather radars in Switzerland taken during intense solar activity in the first seven months of 2014. Section 2 presents the operational Swiss approach and a simplification of the Holleman *et al.* [2,3] method. The available data, as well as the main radar features, are described in Section 3. The results are given in Section 4, together with a possible interpretation of the main features observed.

2. Operational Observations of Solar Spectral Power at the Antenna Feed

A complete description of the automated detection of Sun signatures in polar volume data of weather radar can be found in [1]. A similar approach is currently used at MeteoSwiss: a continuous, non-coherent, noise-like power signal from a non-polarized radio-frequency source is sought at long ranges in consecutive radar bins (75 m long and 1° wide); such a signal is expected to be between 5 and 12 dB stronger than the background noise. Eight-bit, "uncorrected" reflectivity data are used in this study. The pulse repetition frequency and rotation speed are optimized in the operational MeteoSwiss radar scan program as a function of the angle of elevation. Therefore, the number of pulses involved in each 1° half power beam width (HPBW) bin ranges from 28 to 42. In the MeteoSwiss approach, only the last 5 km (60 gates) of each radar bin are averaged for Sun power retrieval. Hence, each reflectivity value corresponding to a Sun hit is the linear average of a few thousand samples. Several tens of hits are found per day, depending on the scan program, latitude and season.

In order to retrieve the solar spectral power, daily radar reflectivity values were analyzed by Holleman *et al.*, in single [2] and dual-polarization [3] receiving chains using the five-parameter linear model fitting procedure described in detail in [1,4]. It is worth noting that a correction factor has been introduced in the recent paper by Huuskonen *et al.* [4] (Equation (3), page 1705) to characterize the residual bias between the true Sun azimuth from ephemeris and the value "read" using the antenna encoder: such a term, which is basically the sine of the Sun zenith angle, is more important in mid-latitude and equatorial regions, where the Sun reaches lower zenith angles. Once the five parameters are derived, for instance by means of the least squares method, the peak solar power that the radar would have received if the beam had hit the Sun's center can be estimated. At MeteoSwiss, the focus of the operational Sun check method is on a relative calibration between horizontal and vertical channels, as well as on an inter-comparison of different radars. Since the focus is not on absolute calibration, the median of the strongest 21 hits in a given day is taken in order to obtain a robust estimate of the Sun's apparent reflectivity. One important implication of this choice is that an accurate assessment of the absolute calibration of the receivers cannot be pursued: such a daily median value is of the order of 2 dB (or more) smaller than the true peak solar power value. However, the method allows an accurate monitoring to be made of the differential bias between the horizontal and vertical channels. Furthermore, assuming that the convolution between the antenna radiation pattern and the Sun's disk will on average lead to approximately the same amount of "underestimation", the stability of radar receiving chains can also be monitored. Section 4 will show that not only is this the case, but also that the monthly variability of the Sun's microwave signals is captured well by the Swiss radars.

3. Description of Acquired Data and Instruments

MeteoSwiss has recently renewed its weather radar network: antenna-mounted, fully-digital receivers have been introduced into the current network; each system is equipped with two orthogonal receiving channels, which are able to measure vertical and horizontal linear polarizations. Three radar systems have already passed the extensive acceptance tests (test readiness review, factory acceptance test, integration and site adaptation test, system equipment acceptance test; see, for instance, [5,6]) and have been operating for more than two years. These three radar sites are: Lema (162 m altitude, near Lugano; Lat 46.04°; Lon 8.83°), Albis (935 m, near Zurich; Lat 46.29°; Lon 8.51°) and Dole (1681 m, near Geneva; Lat 46.43°; Lon 6.10°). From an operational Sun check method viewpoint, probably one of the most relevant aspects to emerge from such acceptance tests was the necessity of a modification of the original calibration unit (CU); updated, new generation calibration units [7] were installed on both the Albis and Lema radars at the end of 2013. Hence, the data shown in this note (from the beginning of 2014 up to mid-August 2014) refer to the new concept CU in two out of three radars.

The solar flux at a 10.7-cm wavelength (S-band) is continuously monitored at the Dominion Radio Astrophysical Observatory (DRAO). This observatory is located near Penticton, British Columbia, Canada, and is characterized by low interference levels at the decimeter and centimeter wavelengths; the quality of the environment is maintained through strong local, provincial and federal protection [8]. Observations of the daily solar flux were started in 1946 and have continued to the present day. The current solar flux at 10.7 cm, $F_{10.7}$, can be obtained from the DRAO observatory website (ftp://ftp.geolab.nrcan.gc.ca/data/solar_flux/daily_flux_values/fluxtable.txt). Such values are corrected for atmospheric attenuation. In particular, they refer to precise measurements acquired three times a day. In summer, this takes place at 17, 20 (local noon) and 23 UTC. The hilly horizon and 50°N latitude cause the Sun to be too low for the first and last timings during the "winter season"; as a consequence, the flux determinations are made at 18, 20 and 22 UTC during "darker months".

The 10.7 cm solar flux measurements can be converted to other wavelengths with some uncertainty (approximately 1 dB according to [8]). This is possible thanks to the remarkable stability of the blackbody spectrum of the slowly varying component of solar activity. However, it should be recalled that the spectral component can be distorted by superimposed contributions from flares, especially for "short" sampling periods, like radar "hits during operational weather radar scanning". In short, on the basis of Equation (13) in Section 5 of [8], the estimated solar flux in solar flux units (sfu) used in this note for the Swiss wavelength (5.5 cm) is:

$$F_{5.5} = 0.7155 \times (F_{10.7} - 64) + 113. \tag{1}$$

If the receiver bandwidth (in Hz), the effective antenna area (in m^2) and the receiver and transmitter losses (dimensionless) are known, then the radar received power can be converted into sfu (see Section 2 in [2] for details); this is also the approach that has been followed in this paper (see the next section). Because of the multiplicative nature of the errors that affect radar measurements, radar-related results are often presented using decibels (dB). In the present case, $s = 10 \log(S/S_0)$ is used, where $S_0 = 1$ sfu $= 10^{-19}$ mW·m^{-2}·Hz^{-1} and $[s] = $ dBsfu.

4. Results and Discussion

4.1. Choice of the Reference Solar Signal and Scores for the Assessment of the Agreement

As stated in the Introduction, the focus of the present analysis is on the relative agreement and not on the absolute calibration of each receiver. In other words, emphasis is on the variation of the error around the mean; it is not on the mean error itself. The error is defined in Sections 4.2 and 4.3 as the ratio between the value measured by the radar receiver and the reference (DRAO measurements); in Section 4.4, the error is the ratio between the horizontal and vertical polarization values. As a consequence, the selected scores emphasize the agreement between oscillations around the mean value. The first selected score is the explained variance, which is the square of the correlation coefficient (multiplied by a hundred). The second score is named "scatter" [9] and is measured in decibels (dB). It has the advantage of being conceived of to be robust, informative and independent of any multiplicative factor that could affect the measurements. The error distribution is expressed as the cumulative weighted contribution to the normalized total amount, which is the reference (y-axis) as a function of the log-transformed estimate/reference ratio (x-axis). Alternatively, it is possible to simply use the standard deviation of the log-transformed radar-to-DRAO solar flux value ratio, as shown, for instance, by Holleman et al. [2].

Figure 1 shows the daily variability of the three series of daily precise measurements acquired in 2014 by DRAO at 10.7 cm and extrapolated to the Swiss radar wavelength using Equation (1): the blue curve is for 20 UTC measured data; the green one is for the 18 (17) UTC data; while the red one is for the 22 (23) UTC data. Because of the Sun's rotation on its axis (variable, as a function of latitude), the number of active regions that enhances the radio emission with respect to the quiet (background) component, as seen from the Earth, varies. The result is an oscillation with an amplitude of approximately 1 dB and a period of ~27 days. In January, there were a few episodes when radiation was outshone; for instance, up to 26.9 dBsfu on 7 January between 17 and 18 UTC. Apart from these few days in January, the three curves are almost indistinguishable. It should be noted that the flux values were log-transformed before being displayed and are shown on the y-axis in dBsfu. Let us now evaluate the agreement between the three series using the above-presented scores: the blue vs. green explained variance is 87%; the blue vs. red one is 98%; and the red vs. green one is 82%.

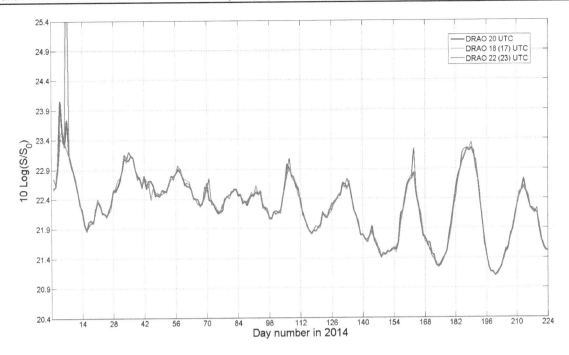

Figure 1. Daily solar flux values measured by the Dominion Radio Astrophysical Observatory (DRAO) in Canada at 10.7 cm and extrapolated to the C-band. Each flux determination takes an hour. Note that $S_0 = 1$ sfu $= 10^{-22}$ W·m^{-2}·Hz^{-1}.

Because of the rapidly varying solar component that consists of bursts (second and minute duration) produced by flares and other transient activity, which caused two extreme variability cases on 4 and 7 January, the explained variance, based on linear values of solar fluxes, is significantly smaller: 66%, 97% and 58%, respectively. As far as the scatter is concerned, the results are listed in Table 1 (the reference measurement is shown along the columns). According to the scatter values, the dataset acquired at 20 UTC again shows better agreement than the other two sets. For this reason, the radar derived data in Section 4.2 will be evaluated against a reference using the 20 UTC dataset.

Table 1. Agreement in terms of scatter (in dB) between the three 1-h long Sun measurements acquired by the DRAO in Canada.

	DRAO 20 UTC	DRAO 18 (17) UTC	DRAO 22 (23) UTC
DRAO 20 UTC	0 dB	0.041 dB	0.044 dB
DRAO 18 (17) UTC	0.040 dB	0 dB	0.056 dB
DRAO 22 (23) UTC	0.044 dB	0.057 dB	0 dB

4.2. Results of the Daily Online Analysis (Horizontal Polarization)

Figure 2 shows the daily analysis of the Sun's signals detected during the operational scan program by the new generation MeteoSwiss meteorological radars operating at the C-band. As in Figure 1, the results are shown from 1 January until 12 August 2014 (Julian Day 224); the retrieved solar flux values in sfu are also transformed to decibels. The solar "monthly" oscillation is clearly visible in all of the radar systems and shows good correlation (see the first line of Table 2) to the reference (DRAO at 20 UTC, as described in the previous section): the Albis radar shows the best agreement with the highest correlation and lowest scatter. A random component (daily fluctuations) seems to affect the radar estimates: a three-day median average (smoothing) is able to make the radar curves more similar to the reference and improve the correlation (see Section 4.3).

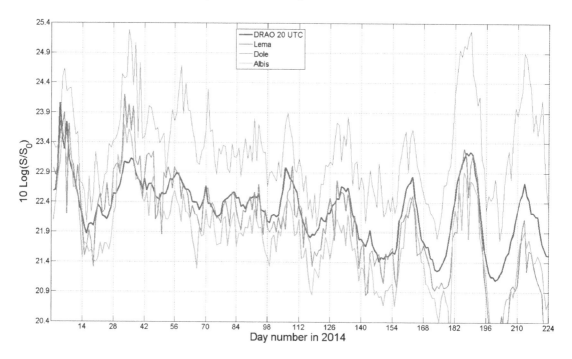

Figure 2. Daily variability of the Sun's emission retrieved from the MeteoSwiss weather radars in Lema, Dole and Albis. The values from DRAO (20 UTC) are plotted for reference purposes.

Table 2. Agreement between the Swiss radar horizontal channel and the DRAO reference (20 UTC).

	Albis Radar	Lema Radar	Dole Radar
Explained variance in %	83%	74%	62%
Scatter in dB	0.28 dB	0.42 dB	0.44 dB
Standard deviation in dB	0.35 dB	0.48 dB	0.48 dB

The same agreement and similar scores are obtained when the 22 (23) and 18 (17) UTC time series are used: our interpretation is that the two selected scores are robust and can be used effectively for a quantitative evaluation of the radar *versus* reference (DRAO) agreement. When using the (less robust) standard deviation of the log-transformed ratios, the Albis radar again shows the best agreement (last line in Table 2). A possible explanation for the better results of the Albis radar could be the use of the stable, new generation calibration unit.

Such values can be compared with those obtained for the Dutch De Bilt radar by Holleman *et al.* [2] in 2005 using DRAO 22 UTC as a reference: the standard deviation is 0.44 dB for the whole year (excluding three days with flares) and is 0.58 dB and 0.44 dB for the first nine months of the year, with (272 days) and without (269 days) the above-mentioned three flares, respectively. As expected, the scatter values are much less sensitive to the three flares: such values are 0.27 dB (272 days) and 0.26 dB (269 days), respectively. Another interesting benchmark is the value shown by

Huuskonen *et al.* [4] for the Finnish Anjalankoski radar over a period of two months at the end of 2011: the standard deviation is as small as 0.24 dB.

Why is the amplitude of the oscillation of the DRAO reference signal smaller than those retrieved by radars? A possible explanation could be related to the transposition formula (Equation (1)) from the S-band to the C-band, which is less accurate when solar activity increases. For instance, applying a slope coefficient of 0.8 instead of 0.7155 in Equation (1) causes a reduction of both the scatter and standard deviation for all three Swiss radars.

4.3. Improvement Obtained When the Random Error Component That Affects Radar Estimates Is Smoothed

The presence of the random error component that affects radar estimates can be pointed out through a simple low-pass (smoothing) filtering. Even a ("short-window") three-day running median is able to significantly reduce daily radar fluctuations, and this results in an improvement in the explained variance and, most of all, in the scatter in dB (see Table 3).

Table 3. Agreement between the DRAO reference (20 UTC) and the Swiss radar horizontal channel after having applied a three-day running median filter.

	Albis Radar	Lema Radar	Dole Radar
Explained variance in %	88%	80%	66%
Scatter in dB	0.05 dB	0.07 dB	0.07 dB
Standard deviation in dB	0.29 dB	0.41 dB	0.43 dB

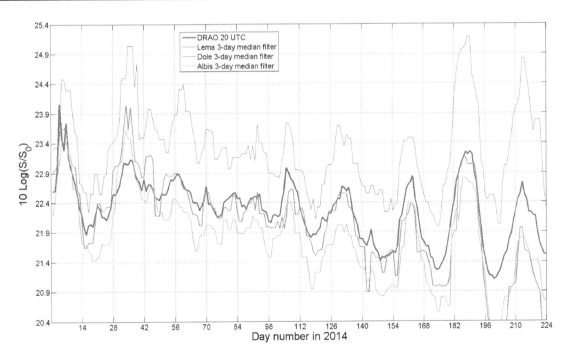

Figure 3. Daily variability of the Sun's emission retrieved from the MeteoSwiss weather radars after having applied a three-day running median to the radar data. The reference measurements acquired by DRAO are also plotted in blue.

From a visual comparison, a fair agreement can be observed between the DRAO reference value and the values retrieved from radar observations after the three-day median filter. Figure 3 shows the corresponding results: the solar rotation, which can be observed in the radio frequency flux during active periods, can clearly be seen in the signatures of all three radar systems.

4.4. Operational Monitoring of Radar Differential Reflectivity Using the Sun

The issue of differential reflectivity calibration is crucial for successful applications of polarimetric weather radar [10]. Offline Sun measurements, in which operational scanning is stopped and the radar antenna is pointing towards the Sun, are generally employed to calibrate the polarimetric receiving chain ([10,11]). Illingworth [12] has shown that Z_{DR} should be estimated within 0.2 dB for improved rainfall estimates based on Z and Z_{DR}. In the present paper, the approach introduced by Holleman *et al.* [3] is followed. The same volume data available from the Swiss weather radars described in Section 4.2 have been analyzed on a daily basis, and the corresponding solar-related results are presented in Table 4.

Table 4. Daily agreement between the horizontal and vertical channel Sun measurements.

	Albis Radar	Lema Radar	Dole Radar
Number of median daily values used	220	204	203
Median value in dB	−0.65 dB	0.17 dB	−0.40 dB
Average value in dB	**−0.66 dB**	**0.18 dB**	**−0.40 dB**
Standard deviation in dB	**0.05 dB**	**0.06 dB**	**0.06 dB**

Twenty values (from 6 June to 25 June) have been discarded for the Lema radar, because of a problem with the calibration of the vertical channel: a residual, positive bias of approximately +0.2 dB is clearly evident. The daily differential reflectivity values from the solar monitoring in fact are consistently positive; only some minor random fluctuations are present, and as a result, the dispersion is as small as 0.06 dB. For the Albis radar, 220 (out of 224) observations are available; the system did not work over two weekends in July. Unfortunately, despite the new-generation calibration unit, a significant negative Bias can be observed: the daily differential reflectivity values from the solar monitoring are systematically negative. However, a positive fact is offered by the limited dispersion: both the standard deviation and spread are 0.05 dB, the smallest value of the three radars. The Dole radar also shows a negative bias of approximately −0.4 dB (21 values have been discarded because of a problem with the calibration of the vertical channel). The dispersion is as small as 0.06 dB.

The limited dispersion found for all three radars is an important result: it underlines the stability of the relative calibration between the horizontal and vertical receivers. It also makes it feasible to plan an adjustment of one channel with respect to the other in order to eventually reach a residual bias as close as possible to the ideal value of 0.0 dB.

5. Summary and Conclusions

The electromagnetic power from the Sun can be used to check the quality of the polarimetric receivers of any weather radar network [9]. In recent years, the Dutch (KNMI) and Finnish (FMI)

Meteorological Institutes have developed an "online" Sun method, in which the Sun's signals are automatically detected in polar volume data generated during operational scan programs. This method has proved to be successful for the daily monitoring of antenna electromagnetic pointing [1], of single polarization radar receiving chain stability [2], as well as for assessing the accuracy of the differential reflectivity of polarimetric weather radars [3]. Accurate estimates of the solar flux have been presented during a quiet period of the roughly 11-year slowly varying radio emission Sun cycle. This note describes results that have recently been obtained during the current, very active Sun period.

As far as the relative agreement between the horizontal polarization receiving chain and the DRAO reference is concerned, Holleman *et al.* [2] presented results for the first 46 days in 2008 for the Dutch De Bilt and Den Helder radars and for 44 days (from 1 July to 13 August 2008) for the Finnish Vimpeli and Luosto radars: the standard deviations of the log-transformed radar-to-DRAO daily ratios resulted in being 0.14, 0.17, 0.16 and 0.20 dB, respectively (quiet Sun period). As far as the MeteoSwiss weather radars are concerned, this note presents results obtained for the first 224 days of 2014. Despite the very active Sun period, satisfactory results have been obtained, as pointed out in Section 4.3 and shown in Figure 3. The Albis radar shows the best performance: the standard deviation of the log-transformed daily ratios is 0.35 dB (0.27 dB in case of the seven-day median filter). The values is 0.48 dB for the Lema and Dole radars (0.40 dB and 0.38 dB, respectively, in the case of the seven-day median filter).

As far as the stability of differential reflectivity is concerned, Holleman *et al.* [3] have presented results for a three-month quiet Sun period (March, April and May) in 2008 for the MeteoFrance radar located in Trappes: they obtained a standard deviation of the daily differential reflectivity that is slightly larger than 0.2 dB. Albis shows the best performance of the Swiss radars for the first seven months of 2014 (active Sun period): the standard deviation of the daily differential reflectivity is as small as 0.05 dB. The Lema and Dole radars also show good results (0.06 dB). It is possible to conclude that the method presented in this note, despite its simplified form, has the potential to routinely monitor dual-polarization weather radar networks, even during periods of intense Sun activity.

Acknowledgments

The authors would like to thank Peter Gölz and Dennis Vollbracht for the stimulating discussions and helpful hints regarding the Sun's observations. Special thanks are due to: Andrea Lombardi for helping the algorithm implementation; Ken Tapping for the kind and helpful suggestions via email in August 2014; Iwan Holleman for having provided the De Bilt radar data for 2005, another active Sun period. The valuable and helpful comments from the three anonymous reviewers are greatly appreciated.

Author Contributions

Marco Gabella, Maurizio Sartori and Marco Boscacci conceived and designed the experiment; Marco Boscacci conceived and programmed the automatic, real-time Sun detection and monitoring algorithm; Urs Germann conceived the score named scatter; Marco Gabella analyzed the data and presented the results; after extensive feedback and comments by Maurizio Sartori and Urs Germann, Marco Gabella wrote the original and revised version of this note.

Conflicts of Interest

The authors declare no conflict of interest.

References

1. Huuskonen, A.; Holleman, I. Determining weather radar antenna pointing using signals detected from the Sun at low antenna elevations. *J. Atmos. Oceanic Technol.* **2007**, *24*, 476–483.

2. Holleman, I.; Huuskonen, A.; Kurri, M.; Beekhuis, H. Operational monitoring of weather radar receiving chain using the Sun. *J. Atmos. Oceanic Technol.* **2010**, *27*, 159–166.

3. Holleman, I.; Huuskonen, A.; Gill, R.; Tabary, P. Operational monitoring of radar differential reflectivity using the Sun. *J. Atmos. Oceanic Technol.* **2010**, *27*, 881–887.

4. Huuskonen, A.; Kurri, M.; Hohti, H.; Beekhuis, H.; Leijnse, H.; Holleman, I. Radar performance monitoring using the angular width of the solar image. *J. Atmos. Oceanic Technol.* **2014**, *31*, 1704–1712.

5. Gabella, M.; Sartori, M.; Progin, O.; Germann, U.; Boscacci, M. An innovative instrumentation for checking electromagnetic performances of operational meteorological radar. In Proceedings of the Sixth European Conference on Radar in Meteorology and Hydrology (ERAD2010), Sibiu, Romania, 6–10 September 2010; pp. 263–269.

6. Gabella, M.; Sartori, M.; Progin, O.; Germann, U. Acceptance tests and monitoring of the next generation polarimetric weather radar network in Switzerland. In Proceedings of the 2013 International Conference on Electromagnetics in Advanced Applications (ICEAA), Torino, Italy, 9–13 September 2013; ISBN: 978-1-4673-5678-9.

7. Vollbracht, D.; Sartori, M.; Gabella, M. Absolute dual-polarization radar calibration: Temperature dependence and stability with focus on antenna-mounted receivers and noise source-generated reference signal. In Proceedings of the 8th European Conference on Radar in Meteorology and Hydrology (ERAD2014), Garmisch-Partenkirchen, Germany, 1–5 September 2014; pp. 91–102.

8. Tapping, K. Antenna Calibration Using the 10.7 cm Solar Flux. Available online: http://www.k5so.com/RadCal_Paper.pdf (accessed on 25 September 2014).

9. Germann, U.; Galli, G.; Boscacci, M.; Bolliger, M.; Gabella, M. Quantitative precipitation estimation in the Alps: where do we stand? In Proceedings of the Third European Conference on Radar in Meteorology and Hydrology (ERAD2004), Visby, Sweden, 6–10 September 2004; pp. 2–6.

10. Ryzhkov, A.V.; Giangrande, S.E.; Melnikov, V.M.; Schuur, T.J. Calibration issues of dual-polarization radar measurements. *J. Atmos. Oceanic Technol.* **2005**, *22*, 1138–1155.

11. Pratte, J.F.; Ferraro, D.G. Automated solar gain calibration. In Proceedings of the 24th Conference on Radar Meteorology, Tallahassee, FL, USA, 27–31 March 1989; pp. 619–622.

12. Illingworth, A. Improved precipitation rates and data quality by using polarimetric measurements. In *Advanced Applications of Weather Radar*; Meischner, P., Ed.; Springer Verlag: Heidelberg, Germany, 2004; Chapter 5, pp. 130–166.

3

National Assessment of Climate Resources for Tourism Seasonality in China Using the Tourism Climate Index

Yan Fang and Jie Yin *

School of Tourism and City Management, Zhejiang Gongshang University, 18 Xuezheng Street, Hangzhou 310018, China; E-Mail: 15757132049@163.com

* Author to whom correspondence should be addressed; E-Mail: rjay9@126.com

Academic Editor: Robert W. Talbot

Abstract: Tourism is a very important industry, and it is deeply affected by climate. This article focuses on the role of climate in tourism seasonality and attempts to assess the impacts of climate resources on China's tourism seasonality by using the Tourism Climate Index (TCI). Seasonal distribution maps of TCI scores indicate that the climates of most regions in China are comfortable for tourists during spring and autumn, while the climate conditions differ greatly in summer and winter, with "excellent", "good", "acceptable" and "unfavorable" existing almost by a latitudinal gradation. The number of good months throughout China varies from zero (the Tibetan Plateau area) to 10 (Yunnan Province), and most localities have five to eight good months. Moreover, all locations in China can be classified as winter peak, summer peak and bi-modal shoulder peak. The results will provide some useful information for tourist destinations, travel agencies, tourism authorities and tourists.

Keywords: tourism seasonality; climate resources; Tourism Climate Index; China

1. Introduction

As tourism becomes one of the largest and fastest growing industries, it plays an extremely important role in promoting national and local economic development all over the world. According to Tourism Highlights (2014 Edition), published by the United Nations World Tourism Organization (UNWTO), international tourist arrivals worldwide have increased to 1.087 billion in 2013, bringing in US$ 1.159 trillion in international tourism receipts [1]. Statistics from World Tourism and Travel (WTTC) show that the direct contribution of travel and tourism to the global economy rose by 3.1% in 2013, accounting for 9.5% of total gross domestic product, one in 11 of the world's jobs, 5.4% of world exports, as well as 4.4% of global investment [2]. Notably, China is a major tourism destination, which has made great achievements in tourism over the past three decades. According to the statistics released by the China National Tourism Administration (CNTA), total tourists, which includes the number of domestic tourist and the international tourist arrivals to China, has reached 3.089 billion in 2012, and tourism income was 2.59 trillion yuan, which increased by 15.2% compared to last year [3]. Currently, China has become the largest supplier of tourist resources and the fourth tourism destination worldwide.

It is widely recognized that weather and climate are critical to the tourism sector worldwide [4–11]. Generally, the relationship among climate, weather and tourism has been examined from the perspective of the geography of tourism and climatology [7]. Both show how climate and weather affect the tourism industry. More specifically, climate and weather have a significant impact on the spatial distribution of tourism centers (e.g., the zone for sun and beach tourism or winter sport tourism), tourism resources (e.g., the pattern of tourism climate resources), tourism seasons (e.g., snow cover for snow-dependent activities), tourism supply and demand (e.g., tour schedules of travel agencies and tourists), *etc.* In other words, climate can be characterized as, on the one hand, an asset for destinations and, on the other hand, an important attraction for tourists [12].

Similarly, tourism seasonality is a major issue, which has been noted by tourism researchers for several decades. Butler gave a widely-recognized definition of tourism seasonality: "the temporal imbalance in the phenomenon of tourism, which may be expressed in terms of dimensions of such elements as numbers of visitors, expenditure of visitors, traffic on highways and other forms of transportation, employment and admissions to attractions" [13]. The causes of seasonality in tourism can be put into two categories: one is natural factors, which relate primarily to the climate of a destination (e.g., temperature, sunshine, rainfall, wind speed, humidity); the other one is institutional factors, which reflect social norms and practices (e.g., the schedule of festivals and holidays, the supply of public and private services) [13–16]. The effect of these two factors is found to depend on the types of tourism destinations. For seaside and mountain resorts, natural environments are more important than institutional causes [17], while institutional causes are major factors for destinations with small contrasts in climate, such as Yunnan Province, China [18].

Most previous studies on China's tourism seasonality have primarily explored the characteristics in some provinces or some types of tourist spots [17–21]. More detailed information, such as the tourism seasonality characteristics of the whole country, especially the spatial pattern of tourism seasonality with the effect of climate in China, has not been examined. When discussing the tourism seasonality of climate resources, one must choose a suitable index to assess tourism climate. Several indices have been developed to evaluate climate environments for tourism, including the Tourism Climate Index (TCI) [22], the Beach Comfort Index (BCI) [23,24], the Climate Index for Tourism (CIT) [25], the Modified Climate Index for Tourism (MCIT) [26] and the Physiologically Equivalent Temperature Index (PET) [27–31]. Obviously, the BCI was designed for evaluating the climate conditions of beach holiday resorts, while the PET was used to reflect the thermal properties of tourism climate. TCI is a composite indicator that captures the climatic elements most relevant for general tourism activities, which has been widely used for Europe [32], North America [6], the Mediterranean [33], northwest

Europe [34], European beaches [35], Australia [36] and at the global level [37]. The TCI integrates three essential climatic facets (thermal, aesthetic and physical) into a single index, and it has widespread applicability, because the required climatological data are commonly available. However, its deficiencies can also be noted as follows: (1) the subjectivity of the TCI (e.g., thresholds, weights) is the central weakness of this scheme; (2) weather-sensitive activities, such as beach tourism, cannot be assessed by the TCI without modification. In order to overcome the known limitations of the TCI, a new index for tourism climate assessment, the Climate Index for Tourism (CIT), was proposed by De Freitas *et al.* [25]. However, the CIT specifically pays attention to "sun, sea and sand" (3S) tourism, not general tourism activities. Additionally, Yu *et al.* developed the Modified Climate Index for Tourism (MCIT) using hourly climatic data, which are not available for the macroscopic analysis of tourism seasonality [26]. Therefore, to make the results easily comparable, Mieczkowski's TCI [22] is employed in this analysis due to its comprehensive nature and universal applicability.

This paper assesses the impact of climate on tourism seasonality in China based on the TCI during the period 1981–2010. The structure of this paper is organized as follows: Section 2 describes the research materials and methodological procedures, including the characteristics of the TCI, classification of the annual tourism climate, data requirements and preparation. Section 3 shows the results and discussion, and conclusions are presented in Section 4.

2. Methodology

2.1. The Tourism Climate Index

The Tourism Climate Index (TCI) was developed by Mieczkowski [22] as a composite measure of tourism climate. Initially, 12 climate variables were identified from previous research related to the climate for tourism to represent the TCI; however, seven climatic variables based on monthly means were finally integrated into the TCI, because of meteorological data limitations. These climatic variables are maximum daily temperature, minimum daily relative humidity, daily temperature, daily relative humidity, total precipitation, hours of sunshine per day and wind speed. Furthermore, all of the climate variables were combined into five sub-indices, which are outlined in Table 1.

Table 1. Sub-indices of Mieczkowski's Tourism Climate Index.

Sub-Index	Monthly Climate Variables	Weight
Daytime comfort (CID)	Maximum daily temperature	40%
	Minimum daily relative humidity	
Daily comfort (CIA)	Mean daily temperature	10%
	Mean daily relative humidity	
Precipitation (P)	Total precipitation	20%
Sunshine (S)	Total hours of sunshine per day	20%
Wind (W)	Average wind speed	10%

According to the relative weightings of the incorporated index components, the TCI formula takes on the following expression:

$$TCI = 2(4CID + CIA + 2P + 2S + W)$$

where CID is the Daytime Comfort Index, CIA is Daily Comfort Index, P is the Precipitation Index, S is the Sunshine Index and W is the Wind Index. By using a standardized rating system, each sub-index takes on an optimal rating of 5, and Mieczkowski suggested a classification scheme for the TCI scores. In order to reflect the distribution of the TCI more directly, a simplified rating system combined with the TCI classification scheme of Mieczkowski is presented [6,36,37] (Table 2).

Table 2. The classification scheme of the Tourism Climate Index (TCI).

TCI Categories of Mieczkowski (1985)	A Simplified Rating System for TCI
Ideal (90–100)	Excellent (80–100)
Excellent (80–89)	Good (60–79)
Very good (70–79)	Acceptable (40–59)
Good (60–69)	Unfavorable (Below 40)
Acceptable (50–59)	
Marginal (40–49)	
Unfavorable (30–39)	
Very unfavorable (20–29)	
Extremely unfavorable (10–19)	
Impossible (Below 9)	

2.2. Annual Tourism Climate Classification

A theory was set up by Scott and McBoyle [6] to classify the annual tourism climate of each destination. All destinations can be divided into six types based on the TCI scores (Table 3): optimal, poor, summer peak, winter peak, bi-modal shoulder peak and dry season peak. In addition, Scott and McBoyle found that monthly hotel or resort accommodation costs followed a similar law of annual tourism climate classification rated with the TCI scores by selecting some locations [6], suggesting that the TCI can represent tourism seasonality (e.g., number of visitors, expenditure of tourists) to some extent.

Table 3. The classification of annual tourism climate by Scott and McBoyle.

Category	Description
Optimal	TCI scores ≥ 80 for each month of the year
Poor	TCI scores < 40 for every month of the year
Summer peak	Summer is the best season in terms of climate conditions
Winter peak	Winter is the best season in terms of climate conditions
Bi-modal shoulder peak	Spring and fall months are more suitable for tourism activities in terms of climate conditions
Dry season peak	The dry season is more conducive to tourist activity in terms of climate conditions

2.3. Data Requirements and Preparation

The major dataset required in the analysis is the standard value of the surface climate dataset in China during 1981 to 2010, which was downloaded from the China Meteorological Data Sharing Service System. With 825 meteorological observation stations in China, this dataset includes station attribute information (e.g., station ID, longitude, latitude) and mean monthly climatology data (e.g., temperature, humidity, precipitation, sunshine, wind speed), which have been used extensively in climate-related research across China [38]. Due to the absence of minimum daily relative humidity in the dataset, as an alternative, we collected this data from the standard value of the surface climate dataset in China during 1971 to 2000.

By matching observation stations in these two datasets and removing incomplete and invalid data through strict quality control procedures, including a thresholds' check, extremes' check, consistency check and completeness check, a total of 658 stations, which were found to be homogeneous without any missing data, were finally retained in the analysis, and the data quality can generally satisfy the research requirements (Figure 1). All of the data required by the TCI were obtained by using MATLAB and then exported to text (.txt) format. Furthermore, the meteorological observation stations and the national administrative boundary map (WGS84 geographic coordinate system) were connected through ArcGIS 10.0 software. The stations are evenly distributed throughout China, except in the western Tibetan Plateau. Taiwan province and some remote islands have been excluded from this study due to a lack of long-term monitoring data series in some parts of these regions.

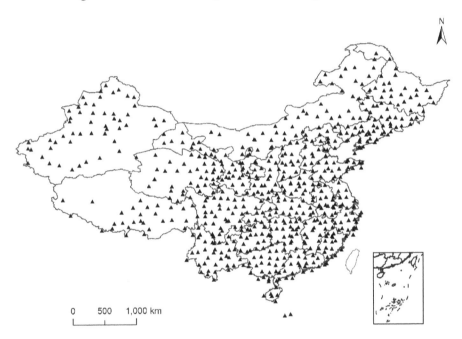

Figure 1. The locations of the 658 meteorological observation stations across China.

3. Results and Discussion

Three sets of analyses were conducted to present the findings and the resulting discussion. First, we calculated the TCI scores for four seasons, while considering the representative characteristic and good stability compared with the TCI scores for each month. Second, the number of good months (TCI $\geq$ 60) is illustrated for each region throughout the year. Third, we classified the TCI values and then described the types of annual tourism seasonality.

3.1. Average Seasonal TCI Scores

Figure 2 presents the TCI values across China for the period 1981–2010 in different seasons. Overall, most of the country has a favorable climate for tourism, such that tourists may visit anytime with many choices available. During spring, most parts of northern and northwestern China are perceived as having good conditions ($60 \leq TCI \leq 79$). The northeastern, eastern and southern regions are classified as having acceptable conditions ($40 \leq TCI \leq 59$). The former areas have higher TCI scores, because of the moderate temperatures and relative dry weather in spring, which are called the spring drought [39]; conversely, the low temperature is responsible for the low scores of the northeast and Tibetan Plateau, while a humid climate with abundant precipitation leads to those for the southern regions. The average TCI map during autumn shows the same categories with good and acceptable conditions during spring, and the essential difference between these two seasons is that the regions characterized as "good" stretch to eastern and southern China during autumn, which can be generally attributed to the decreasing humidity. In summer and winter, China obviously exhibits a latitudinal gradation of tourism climatic attractiveness. Due to the cool climate in summer, the excellent ($TCI \geq 80$) and good conditions ($60 \leq TCI \leq 79$) mainly occur in the north of the Qinling Mountains and Huaihe River. In winter, the regions considered as "good" are located in the south of the Five Ridges and Yunnan Province, and regions considered as "unfavorable" are located in the northeast, northwest areas and Inner Mongolia; the other parts are generally categorized as "acceptable". Whether a region has an attractive tourism climate for tourists in summer and winter, in general, depends on the temperature, which is significantly influenced by the latitude. However, at the local scale, this is likely influenced by the topography and environment compared with the regions at a similar latitude. For example, Shandong Peninsula is affected by a maritime climate and has a cooler summer than other areas at the same latitude. It should be borne in mind that the results in Figure 2 probably cannot reflect the actual tourism activities in each region, because some other factors, such as institutional factors and cultural and historical factors (such as historical sites and scenic spots), have not been considered in the study.

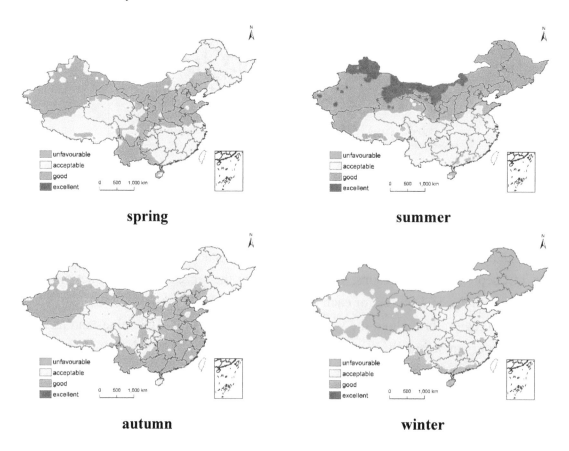

Figure 2. TCI values for China during the four seasons (1981–2010).

A comparison of the average seasonal TCI maps during the four seasons shows that most of China is classified as having "good" or "acceptable" conditions all year, but regions with "excellent" or "unfavorable" ranks only exist in summer and winter, respectively. There is no doubt that the map based on the TCI approach, which describes tourist climatic comfort, performs quite well as an assessment of climate resources for tourism seasonality and a predictor of visitation. Although the map cannot reflect the actual number of tourists due to several causes of tourism seasonality, it helps tourism authorities think about what they should do to increase the number of travelers under an uncomfortable climate or whether they perform satisfactorily with respect to the attraction of visitors in a comfortable climate. Taking Harbin City (the capital of Heilongjiang Province) as an example, it has spared no expense to develop ice-snow tourism in winter (e.g., Harbin International Ice and Snow Festival), transforming the disadvantage of the cold climate into an advantage successfully.

3.2. Numbers of Good Months

The number of good months (TCI ≥ 60) in a whole year, as a picture of current tourism comfort levels throughout China, is presented in Figure 3. In terms of climate resources, China shows notable variation in the number of good months, varying from zero to 10. The regions with the lowest number of good months (less than four), include the Tibetan Plateau area (e.g., eastern Tibet, southern Qinghai) and the most part of southern China (e.g., Jiangxi Province, Hunan Province). It should be pointed out, however, that these areas with less than four good months experience about nine months of acceptable conditions, indicating that although the climate resource is not an advantage in these areas, it would not restrict the development of tourism. Specifically, eastern Tibet and southern Qinghai are classified as having unfavorable conditions only in the winter, while the central areas of China are "good" in fall (Figure 2). As a special alpine-cold zone, the Tibetan Plateau area has low temperatures all year round, because of its high elevation, resulting in low scores in CID and CIA. However, less precipitation, as well as more sunshine duration lead to acceptable conditions during most of the year.

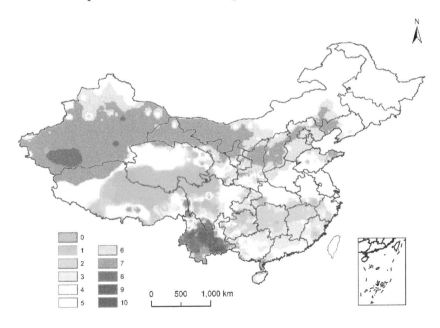

Figure 3. The number of good travel months in China (1981–2010).

The most attractive area with more than eight good months is Yunnan Province. As is well known, the climate of Yunnan Province is mild without a hot summer or a bleak winter because of its special geographic location (low latitude plateau) and the general air circulations. In terms of general air circulations, Yunnan province is controlled by a humid maritime airstream from May to October, while it is controlled by a dry-warm continental airstream from November to April, both of which tend to result in long-term pleasant climates. For example, the suitable period for travelling in Lijiang (a small city in Yunnan Province) is about nine months, from March to November [40]. Moreover, the climatic attractiveness of Lijiang has improved because of the warming-drying trend since 2000. The number of good months in the majority of north China is between five and eight, suggesting that the climates of most areas are comfortable for traveling for about half of the year.

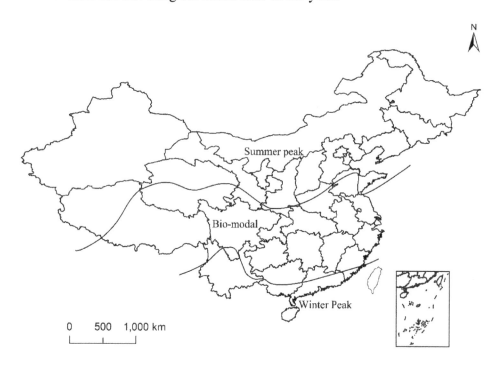

Figure 4. Spatial distribution of annual tourism seasonality (1981–2010).

3.3. Annual Tourism Seasonality Types

Based on the classification scheme of the annual tourism climate suggested by Scott and McBoyle [6] (Table 3), three types, namely the winter peak, summer peak and bi-modal shoulder peak, are identified in China. The spatial distribution of annual tourism seasonality types affected by climate throughout China is presented in Figure 4. Overall, the spatial distribution across China generally shows a latitudinal gradation, with a summer peak, a bi-modal shoulder peak and a winter peak running from north to south. The dividing line between the summer peak and bi-modal shoulder peak approximately follows the boundary of the Qinling Mountains and Huaihe River. As we know, the Qinling Mountains and Huaihe River are important geographic boundaries dividing the north and south parts of China. Therefore, regions to the north of the Qinling Mountains and Huaihe River can be classified as summer peak, except eastern Tibet and southern Qinghai. Regions to the south of the Qinling Mountains and Huaihe River can further be divided into two classes: the southernmost areas of China are winter peak destinations, including southern Guangdong, Guangxi, Yunnan and all areas of Hainan Province; the other areas are bi-modal shoulder peak destinations. As previously suggested (Section 3.2), Yunnan Province has higher TCI scores compared with regions at a similar latitude, leading to the dividing line between bi-modal and summer peak moving southward sharply toward the left.

This classification cannot apply to all destinations perfectly. The TCI values of Yunnan Province are in the range from 60 to 79 all year round; although ratings in the 80s occur in early spring, and 50s occur in late summer. The distribution with high scores lasting for nearly one year would have been labelled "optimal", but its TCI scores are not high enough to be over 80 every month. Yunnan Province, according to the calculated result (Section 3.1), is finally classified as winter peak, because it is distinctly superior to most regions in this season. Poor destinations, moreover, do not exist in China.

4. Conclusions

This paper presents an empirical analysis of the climate impact on tourism seasonality in China during the period of 1981–2010, by combining high quality national meteorological datasets with GIS. The results reveal that the climates of most regions in China are comfortable for tourists during spring and autumn. In summer and winter, the spatial distribution of tourism climates shows a latitudinal gradation. Furthermore, the study suggests that the number of good months throughout China varies from zero to 10, and the most attractive area, with high levels of tourism comfort, is Yunnan Province, whereas the Tibetan Plateau area has the least attractiveness. All regions in China can be divided into three types of annual tourism climate: winter peak, summer peak and bi-modal shoulder peak. The importance of climate resources for tourism seasonality, as well as the spatial distribution of tourism seasonality affected by climate are underlined by analytical results.

All findings of this study can be used as a reference for traveling, planning of destinations, adaptions for reducing seasonality, *etc.* Based on the TCI, travel agencies can create different tour schedules in different seasons. Additionally, a different TCI can be utilized to design different tourism products and even marketing programs. Tourism authorities may make proper readjustments to reduce seasonality by considering climatic attractiveness. What is more, the results of this paper can also be used to promote tourism destinations by local governments (e.g., improvement of existing destinations, development of new destinations). However, the potential effects of climate change will be profound, with temperatures rising and more extreme weather occurring. For example, global surface temperature has risen by 0.8 °C over the last 100 years based on land and marine data [41]. In order to arrive at more robust predictions for future adaptation, additional research is required to compare the results with the actual tourist number and to measure the spatial-temporal distribution of climate resources for tourism seasonality responding to the projected climate change, assessing the effects of climate change in different regional tourism industries, as well as how to deal with these changes flexibly.

Acknowledgments

This research was supported by the National Natural Science Foundation in China (Grant No. 41201550), the Humanities and Social Science Project of the Education Ministry (Grant No. 12YJCZH257) and the Key Project of the Young Talent Foundation from Zhejiang Gongshang University (Grant No. QZ13-2), the Humanities and Social Science Key Research Base of Zhejiang

Province (Applied Economics at Zhejiang Gongshang University) (JYTyyjj20130105) and the Graduate Innovation Foundation of Zhejiang Gongshang University (1040XJ1513027).

Author Contributions

Yan Fang and Jie Yin conceived the study. Yan Fang gathered the weather data, carried out the caculations and analysis of the results. Jie Yin designed the framework, edited and revised the manuscript.

Conflicts of Interest

The authors declare no conflict of interest.

References

1. UNWTO. UNWTO Tourism Highlights. Available online: http://dtxtq4w60xqpw.cloudfront.net/ sites/all/files/pdf/unwto_highlights14_en.pdf (accessed on 3 July 2014).
2. WTTC. Economic Impact of Travel & Tourism 2014 Annual Update: Summary. Available online: http://www.veilleinfotourisme.fr/medias/fichier/economic-impact-summary-2014-publie-le-03032 0142ppa4-final_1395222502243-pdf (accessed on 3 July 2014).
3. CNTA. Statistical Bulletin of Tourism Industry in China on 2012. Available online: http://www.cnta.gov.cn/html/2013-9/2013-9-12-%7B@hur%7D-39-08306.html (accessed on 7 July 2014). (In Chinese)
4. Smith, K. The influence of weather and climate on recreation and tourism. *Weather* **1993**, *48*, 398–404.
5. De Freitas, C.R. Tourism climatology: Evaluating environmental information for decision making and business planning in the recreation and tourism sector. *Int. J. Biometeorol.* **2003**, *48*, 45–54.
6. Scott, D.; McBoyle, G.; Schwartzentruber, M. Climate change and the distribution of climatic resources for tourism in North America. *Clim. Res.* **2004**, *27*, 105–117.
7. Gomez-Martín, B. Weather, climate and tourism. *Ann. Tour. Res.* **2005**, *32*, 571–591.
8. Berrittella, M.; Bigano, A.; Roson, R.; Tol, R.S.J. A general equilibrium analysis of climate change impacts on tourism. *Tour. Manag.* **2006**, *27*, 913–924.
9. Goh, C. Exploring impact of climate on tourism demand. *Ann. Tour. Res.* **2012**, *39*, 1859–1883.
10. Becken, S. A review of tourism and climate change as an evolving knowledge domain. *Tour. Manag. Perspect.* **2013**, *6*, 53–62.
11. Rosselló, J. How to evaluate the effects of climate change on tourism. *Tour. Manag.* **2014**, *42*, 334–340.
12. Day, J.; Chin, N.; Sydnor, S.; Cherkauer, K. Weather, climate, and tourism performance: A quantitative analysis. *Tour. Manag. Perspect.* **2013**, *5*, 51–56.
13. Butler, R.W. Seasonality in tourism: Issues and implication. In *Tourism: A State of the Art*; Seaton, A.V., Ed.; Wiley: Chichester, UK, 1994; pp. 332–339.
14. Bar-On, R.V. *Seasonality in Tourism*; The Economic Intelligence Unit: London, UK, 1975.
15. Bar-On, R.V. The measurement of seasonality and its economic impacts. *Tour. Econ.* **1999**, *5*, 437–458.
16. Hartmann, R. Tourism, seasonality and social change. *Leis. Stud.* **1986**, *5*, 25–33.

17. Lu, L.; Xuan, G.F.; Zhang, J.H.; Yang, X.Z.; Wang, D.G. An approach to seasonality of tourist flows between coastland resorts and mountain resorts: Examples of Sanya, Beihai, Mt. Putuo, Mt. Huangshan and Mt. Jiuhua. *Acta Geogr. Sin.* **2002**, *57*, 731–740. (In Chinese)

18. Li, Y. Analysis of the causes of Yunnan tourism seasonality and the mechanism of the impact. *Tour. Forum* **2009**, *2*, 573–577. (In Chinese)

19. Liang, Z.X.; Bao, J.G. A seasonal study on tourist flows in theme parks during golden weeks—A case of theme parks in Shenzhen overseas Chinese town. *Tour. Tribune* **2012**, *27*, 58–65. (In Chinese)

20. Ma, S.H.; Dai, L.L.; Wu, B.H. Analysis on the features and causes of seasonality in rural tourism: A case study of Beijing suburbs. *Prog. Geogr.* **2012**, *31*, 817–824. (In Chinese)

21. Chen, Y.; Tian, L. A study on the characteristics of Hainan inbound tourism seasonality and regulation measures. *Hum. Geogr.* **2013**, *28*, 140–143. (In Chinese)

22. Mieczkowski, Z. The tourism climatic index: A method of evaluating world climates for tourism. *Can. Geogr.* **1985**, *29*, 220–233.

23. Becker, S. Beach comfort index—A new approach to evaluate the thermal conditions of beach holiday resorts using a South African example. *GeoJournal* **1998**, *44*, 297–307.

24. Morgan, R.; Gatell, E.; Junyent, R.; Micallef, A.; Ozhan, E.; Williams, A. An improved user-based beach climate index. *J. Coast. Conserv.* **2000**, *6*, 41–50.

25. De Freitas, C.R.; Scott, D.; McBoyle, G. A second generation climate index for tourism (CIT): A specification and verification. *Int. J. Biometeorol.* **2008**, *52*, 399–407.

26. Yu, G.; Schwartz, Z.; Walsh, J.E. A weather-resolving index for assessing the impact of climate change on tourism related climate resources. *Clim. Chang.* **2009**, *95*, 551–573.

27. Matzarakis, A.; Mayer, H.; Iziomon, M.G. Applications of a universal thermal index: Physiological equivalent temperature. *Int. J. Biometeorol.* **1999**, *43*, 76–84.

28. Lin, T.P.; Matzarakis, A. Tourism climate information based on human thermal perception in Taiwan and Eastern China. *Tour. Manag.* **2011**, *32*, 492–500.

29. Farajzadeh, H.; Matzarakis, A. Evaluation of thermal comfort conditions in Ourmieh Lake, Iran. *Theor. Appl. Climatol.* **2012**, *107*, 451–459.

30. Matzarakis, A.; Rammelberg, J.; Junk, J. Assessment of thermal bioclimate and tourism climate potential for central Europe—The example of Luxembourg. *Theor. Appl. Climatol.* **2013**, *114*, 193–202.

31. Li, R.; Chi, X.L. Thermal comfort and tourism climate changes in the Qinghai–Tibet Plateau in the last 50 years. *Theor. Appl. Climatol.* **2014**, *117*, 613–624.

32. Rotmans, J.; Hulme, M.; Downing, T.E. Climate change implications for Europe. *Glob. Environ. Chang.* **1994**, *4*, 97–124.

33. Amelung, B.; Viner, D. Mediterranean tourism: Exploring the future with the tourism climatic Index. *J. Sustain. Tour.* **2006**, *14*, 349–366.

34. Nicholls, S.; Amelung, B. Climate change and tourism in northwestern Europe: Impacts and adaptation. *Tour. Anal.* **2008**, *13*, 21–31.

35. Moreno, A.; Amelung, B. Climate change and tourist comfort on Europe's beaches in summer: A reassessment. *Coast. Manag.* **2009**, *37*, 550–568.

36. Amelung, B.; Nicholls, S. Implications of climate change for tourism in Australia. *Tour. Manag.* **2014**, *41*, 228–244.

37. Amelung, B.; Nicholls, S.; Viner, D. Implications of global climate change for tourism flows and seasonality. *J. Travel Res.* **2007**, *45*, 285–296.

38. Yin, Z.E.; Yin, J.; Zhang, X.W. Multi-scenario-based hazard analysis of high temperature extremes experienced in China during 1951–2010. *J. Geogr. Sci.* **2013**, *23*, 436–446.

39. Mu, W.B.; Yu, F.L.; Xie, Y.B.; Liu, J.; Li, C.Z.; Zhao, N.N. The copula function-based probability characteristics analysis on seasonal drought & flood combination events on the North China Plain. *Atmosphere* **2014**, *5*, 847–869.

40. Cao, W.H.; He, Y.Q.; Li, Z.S.; Wang, S.X.; Wang, C.F. Evaluation of the tourism climate comfort index in Lijiang city, Yunnan. *J. Glaciol. Geocryol.* **2012**, *34*, 201–206. (In Chinese)

41. Efthymiadis, D.A.; Jones, P.D. Assessment of maximum possible urbanization influences on land temperature data by comparison of land and marine data around coasts. *Atmosphere* **2010**, *1*, 51–61.

4

Variations in PM_{10}, $PM_{2.5}$ and $PM_{1.0}$ in an Urban Area of the Sichuan Basin and Their Relation to Meteorological Factors

Yang Li [1,*], Quanliang Chen [2], Hujia Zhao [3], Lin Wang [2] and Ran Tao [2]

[1] Center for Atmosphere Watch and Service, Meteorological Observation Center, China Meteorological Administration, Beijing 100081, China

[2] Plateau Atmospheric and Environment Key Laboratory of Sichuan Province, College of Atmospheric Sciences, Chengdu University of Information Technology, Chengdu 610225, China; E-Mails: chenql@cuit.edu.cn (Q.C.); Lynn23@163.com (L.W.); Taorancuit@163.com (R.T.)

[3] Institute of Atmospheric Environment, China Meteorological Administration, Shenyang 110016, China; E-Mail: tjzhj4659@sina.com

* Author to whom correspondence should be addressed; E-Mail: liyang@cma.gov.cn

Academic Editors: Junji Cao, Ru-Jin Huang and Guohui Li

Abstract: Daily average monitoring data for PM_{10}, $PM_{2.5}$ and $PM_{1.0}$ and meteorological parameters at Chengdu from 2009 to 2011 are analyzed using statistical methods to replicate the effect of urban air pollution in Chengdu metropolitan region of the Sichuan Basin. The temporal distribution of, and correlation between, PM_{10}, $PM_{2.5}$ and $PM_{1.0}$ particles are analyzed. Additionally, the relationships between particulate matter (PM) and certain meteorological parameters are studied. The results show that variations in the average mass concentrations of PM_{10}, $PM_{2.5}$ and $PM_{1.0}$ generally have the same V-shaped distributions (except for April), with peak/trough values for PM average mass concentrations appearing in January/September, respectively. From 2009 to 2011, the inter-annual average mass concentrations of PM_{10}, $PM_{2.5}$ and $PM_{1.0}$ fall year on year. The correlation coefficients of daily concentrations of PM_{10} with $PM_{2.5}$, PM_{10} with $PM_{1.0}$, and $PM_{2.5}$ with $PM_{1.0}$ were high, reaching 0.91, 0.83 and 0.98, respectively. In addition, the average ratios of $PM_{2.5}/PM_{10}$, $PM_{1.0}/PM_{10}$ and $PM_{1.0}/PM_{2.5}$ were 85%, 78% and 92%, respectively. From this, fine PM is determined to be the principal pollutant in the Chengdu region. Except for averaged air pressure values, negative correlations exist between other meteorological parameters and PM. Temperature and air pressure influenced the transport and accumulation of PM by affecting convection. Winds promoted PM dispersion. Precipitation not only accelerated the deposition of wet PM, but also inhibited surface dust transport. There was an obvious correlation between PM and visibility; the most important cause of visibility degradation was due to the light extinction of aerosol particles.

Keywords: PM_{10}; $PM_{2.5}$; $PM_{1.0}$; meteorological parameters; Sichuan Basin

1. Introduction

Atmospheric particulate matter (PM) consists of minuscule particles of solid or liquid matter, with diameters ranging from 0.001 μm to 100 μm. The time for which PM is suspended in the atmosphere ranges from a few hours to a few weeks. The smaller a particle is, the longer it will stay in the air. Atmospheric particles can affect the climate by both direct and indirect radiative forcing [1], and serve as cloud condensation or ice nuclei modifying the microphysical characters of clouds [2,3]. Atmospheric PM concentrations exceeding guidelines may cause poor visibility [4,5] and induce the regional hazy weather frequently seen in China [6–8]. In addition, fine particles also damage respiratory and cardiovascular health, and even DNA [9].

There is a nationwide research focus on PM as the principal pollutant in China, including its physical, chemical and optical properties [10–12]. Most studies have focused on northern China [13–15], the Yangtze River delta [16,17], and the Pearl River delta [18]. As a measure of this concern, the latest ambient air quality standards published in 2012 (GB3095–2012) gave $PM_{2.5}$ values for the first time [19].

The Sichuan Basin is located east of the Tibetan Plateau, with the Qin Mountains to the north and the Yunnan-Guizhou Plateau to the south. Due to the particular topography of the basin, average wind speeds in Sichuan are low. Sichuan is a wet, foggy area and mass concentrations of atmospheric particles are therefore always high there [20,21]. The 2009–2011 Environmental Conditions Bulletin (Sichuan Environment Condition Communique) [22–24] from 2009 to 2011 issued by the Department of Environmental Protection of Sichuan Province showed that the average mass concentration of inhalable particles (PM_{10}) in Sichuan was greater than 60 μg/m³ for each of these three years. Although these values were in the second highest national pollution band, they do not fully reflect the PM levels resulting from the rapid economic development that has occurred in Sichuan over recent years.

Chengdu, in the central Sichuan Plain, is the provincial capital of Sichuan. Its present development and future economic trajectory is ideal for representing PM pollution in the Sichuan Basin [25–27]. The Chengdu Environmental Quality Gazette (2009, 2010, 2011, 2012) [28–31] shows that mean annual concentrations of inhalable particles (PM_{10}) in Chengdu exceeded 100 μg/m³, and that daily rates exceeded the set standards by more than 14% (more than 24% regarding to the latest standards issued in 2012). Tao *et al.* [32] found that average mass concentrations of $PM_{2.5}$ reached 165 μg/m³ from April 2009 to January 2010 at the Chengdu Institute of Plateau Meteorology. Based on the average daily monitored data for PM_{10}, $PM_{2.5}$ and $PM_{1.0}$ at Mt. Chengdu from January 2009 to December 2011, this paper analyzes the correlations between PM_{10}, $PM_{2.5}$ and $PM_{1.0}$ and their temporal distribution characteristics. Additionally, the relations between PM and meteorological factors are studied. The results of this study could be used as a reference point for analyzing the pollution sources and composition of atmospheric particulates in urban areas of the Sichuan Basin as well as formulating measures to counteract them. Subsequent observations of variations in mass concentrations of atmospheric PM could then highlight any deficiencies in pollution control.

2. Methodology

PM_{10}, $PM_{2.5}$ and $PM_{1.0}$ data were collected from a building rooftop next to the Chengdu City Meteorological Bureau (104°02′E, 30°39′N; 587.0 m), 91 m above the ground, from 2009 to 2011. The PM monitoring station is located in the west urban area of Chengdu city. Aerosol samples collected from a 91 m-tall building. There are few factories around. The sampling height is above the urban canopy without much local pollution disturbance. Thus the sampling could be representative the pollution level of Chengdu region. Chengdu region belongs to the subtropical monsoon climate. The average annual temperature is around 16.5 degrees centigrade with an average annual rainfall of 1124.6 mm. The wind speed is low with the average annual wind speed is 1–1.5 m/s. The annual sunshine averages 1042–1412 h. The data were measured using a Grimm-180 device (German online PM monitor). This instrument uses an optical method to measure PM concentration of PM_{10}, $PM_{2.5}$, and $PM_{1.0}$. The measuring range is 1–1500 $\mu g/m^3$ with an accuracy of ±2%. The instrument monitored for 24 h at a time, and transferred data once every five minutes. The measuring principle of Grimm-180 is the scattering light measurement of the single particles, where a semiconductor laser serves as light source. The 90° scattered light with an opening angle of about 60° is being lead via a mirror onto a receiver diode. When particles cross the laser beam they emit a light pulse. This electric signal of the diode will be classified into 31 different size channels after an adequate amplification. This enables a size determination of the particles and establishes also a weighting curve for PM_{10} and $PM_{2.5}$. The sample air is being volume controlled sucked through the optical measurement chamber and a fine filter. The pump also provides the rinsing air, which is generated via the pumps exhaust air and a subsequent fine filter. It is being held constant by a rinsing air controller. The rinsing air prevents the laser optics and measurement chamber from pollution and furthermore being used as particle-free reference air during the self-test. During the January 2009 to December 2011 period, 833 valid samples were acquired (excluding calibration instrument failures, missing data, abnormal data and instrument calibration times). The meteorological data, from the Chengdu Meteorology Observatory (104°01′E, 30°40′N; 507.3 m), included daily wind speed, air pressure, temperature, precipitation, relative humidity (RH) and visibility from 2009 to 2011. The Chengdu Meteorology Observatory is located in the west suburban region of Chengdu. And it is surrounded by the farmland and 15 km west from the PM monitoring station. Chengdu Meteorology Observatory is a national principal station and the observations are used to present the meteorological features of Chengdu region.

A linear correlation method was used to analyze the correlation between PM_{10} and $PM_{2.5}$, and between PM_{10} and $PM_{1.0}$. The total average concentrations of these three particle sizes were calculated using the sum of their daily average mass concentrations divided by the sum of valid days.

3. Results and Discussion

3.1. Monthly, Seasonal and Inter-Annual Variations in Mass Concentrations of PM_{10}, $PM_{2.5}$ and $PM_{1.0}$

Variations in the monthly, seasonal and inter-annual average mass concentrations of PM_{10}, $PM_{2.5}$ and $PM_{1.0}$ are shown in Figure 1. Figure 1a presents the monthly average mass concentrations of PM_{10}, $PM_{2.5}$ and $PM_{1.0}$. Trends for these three PM sizes are similar, presenting a concave parabolic shape for January to April, and April to December. Peak monthly values for average mass concentrations of PM_{10}, $PM_{2.5}$ and $PM_{1.0}$ appeared in January, reaching 118 μg/m³, 101 μg/m³ and 93 μg/m³, respectively. The lowest values were in September and reached 45 μg/m³, 42 μg/m³ and 39 μg/m³, respectively. The highest monthly average concentrations were 2.64 times (PM_{10}), 2.42 times ($PM_{2.5}$) and 2.37 ($PM_{1.0}$) times the lowest monthly average concentrations, respectively. Concentrations of $PM_{2.5}$ in December and January significantly exceeded daily average PM_{10} and $PM_{2.5}$ concentrations by 150 μg/m³ and 75 μg/m³ as defined by the ambient air quality standard GB3095-2012, indicating $PM_{2.5}$ pollution in the Chengdu urban area is severe during the winter.

Figure 1b shows clear variations in the seasonal average mass concentrations of PM_{10}, $PM_{2.5}$ and $PM_{1.0}$. The seasons were demarcated as spring (March to May), summer (June to August), fall (September to November) and winter (December to February). Seasonal variations evidenced a concave parabola shape. PM_{10}, $PM_{2.5}$ and $PM_{1.0}$ maxima appeared in winter, reaching 98 μg/m³, 83 μg/m³ and 77 μg/m³, respectively. Minimum PM_{10}, $PM_{2.5}$ and $PM_{1.0}$ values of 49 μg/m³, 43 μg/m³ and 40 μg/m³, respectively, occurred during the summer.

The higher concentrations of PM_{10}, $PM_{2.5}$ and $PM_{1.0}$ in December and January are probably related to geographic factors. Tao *et al.* [25] showed that that the higher PM concentrations during winter in Sichuan Basin are due to the lower mixing height and low wind speeds. Additionally, the sampling point was located at a busy city intersection with high levels of vehicle exhaust emissions: Chen *et al.* [33] states that lower wet particulate deposition is the major cause of pollution in winter. PM concentrations increased in the spring, most likely because Chengdu is occasionally affected by northerly sand and dust storms. Due to the larger size of dust particles, PM_{10} concentrations increased significantly, contrary to fine PM concentrations. In the summer there are more intensive convective air currents and more rainfall, so PM is more effectively removed, leading to minimum PM concentrations. The mixing ventilation coefficient is larger, so contaminant dispersion is rapid. In addition, rainwater aids the removal of coarse particles, but has less effect on fine particles [34]. This could account for $PM_{2.5}$ concentrations being higher in summer. In September, atmospheric RH decreases and the hygroscopic growth of particles is correspondingly inhibited, a contributory factor for the minimum PM concentration. Conversely, owing to the influence of fall rain in West China after September, atmospheric RH increases and the hygroscopic growth of particles is strengthened, leading to a recovery in mass concentration values for all three PM sizes [35].

Figure 1c shows that the annual average mass concentrations of PM_{10}, $PM_{2.5}$ and $PM_{1.0}$ declined significantly from 2009 to 2011. This could be attributed to the efficacy of pollution control measures introduced in Chengdu, such as limiting the number of cars on the road, closing the worst-polluting factories and banning straw-burning [32].

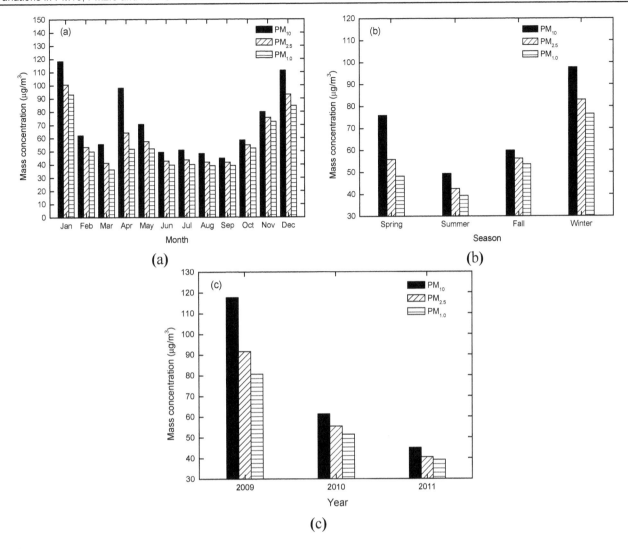

Figure 1. Monthly (**a**), seasonal (**b**) and inter-annual (**c**) average mass concentration variations of PM$_{10}$, PM$_{2.5}$ and PM$_{1.0}$ values.

3.2. Relations between PM$_{10}$, PM$_{2.5}$ and PM$_{1.0}$

The relations between PM$_{10}$ and PM$_{2.5}$, PM$_{10}$ and PM$_{1.0}$, and PM$_{2.5}$ and PM$_{1.0}$ are shown in Figure 2a–c, respectively. The correlation coefficients between them were 0.92, 0.84 and 0.98, meaning the correlations between PM$_{10}$, PM$_{2.5}$ and PM$_{1.0}$ were significant. This result is consistent with previous findings by Deng *et al.* [36]. The ratios of PM$_{2.5}$ to PM$_{10}$, PM$_{1.0}$ to PM$_{10}$, and PM$_{1.0}$ to PM$_{2.5}$ are illustrated in Figure 3. The average PM$_{2.5}$/PM$_{10}$, PM$_{1.0}$/PM$_{10}$ and PM$_{1.0}$/PM$_{2.5}$ values were 85%, 78% and 92%, respectively, and ranged between 74%–94%, 64%–89% and 87%–95%, respectively. The proportions of PM$_{2.5}$ contained within PM$_{10}$ were quite high, further proving that fine particles were the greatest contributor to PM air pollution in Chengdu. The average PM$_{2.5}$/PM$_{10}$ and PM$_{1.0}$/PM$_{10}$ ratios were 72% and 61% in spring, lower than in other seasons. Northerly dust and sandstorms may be a possible cause.

3.3. The Relation between Concentrations of PM$_{10}$, PM$_{2.5}$ and PM$_{1.0}$ and Meteorological Factors

Previous studies have indicated that atmospheric particle pollution is influenced by both the emissions of particles and meteorological factors [37–39]. When particle emissions in urban areas are stable, atmospheric PM pollution is principally caused by meteorological factors, such as the dispersion and transport by wind, and an acceleration of the scavenging by precipitation, of atmospheric PM. Additionally, photochemical reactions depend upon favorable temperature, RH, solar radiation and other conditions. The accumulation and transport of particles is closely related to the synoptic system and atmospheric circulation.

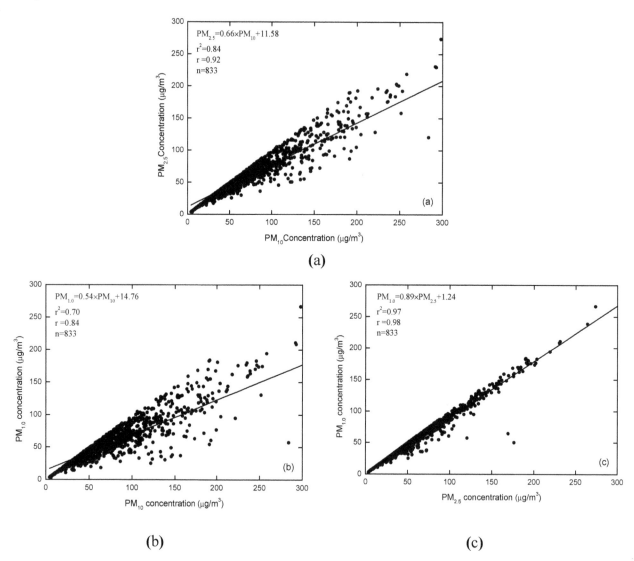

Figure 2. Correlation between average mass concentrations of PM_{10}, $PM_{2.5}$ and $PM_{1.0}$: (**a**) PM_{10} *vs.* $PM_{2.5}$, (**b**) PM_{10} *vs.* $PM_{1.0}$, (**c**) $PM_{2.5}$ *vs.* $PM_{1.0}$.

Table 1 shows that mass concentration of atmospheric PM has a positive relation with air pressure and negative relations with other meteorological factors. The relation between mass concentration of atmospheric PM and visibility is the clearest. The correlations between meteorological factors of air pressure, temperature, precipitation and atmospheric PM pollution are also obvious. However, the correlations between wind speed, RH and PM concentration are not a simple linear one. For example, RH displays no clear correlation with PM concentration.

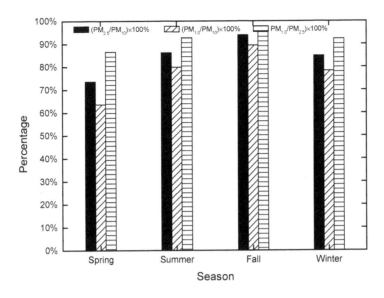

Figure 3. $PM_{2.5}/PM_{10}$; $PM_{1.0}/PM_{2.5}$; and $PM_{10}/PM_{2.5}$ ratios.

Table 1. Correlation coefficients between mass concentrations of PM_{10}, $PM_{2.5}$ and $PM_{1.0}$ and average meteorological factors: wind speed, air pressure, air temperature, precipitation, RH and visibility.

Correlation Coefficient	PM_{10}	$PM_{2.5}$	$PM_{1.0}$
Average wind speed	−0.35	−0.56	−0.62 *
Average air pressure	0.71 *	0.77 **	0.77 **
Average air temperature	−0.73 **	−0.75 **	−0.75 **
Monthly precipitation	−0.59 *	−0.58 *	−0.57
Average RH	−0.25	−0.06	−0.00
Visibility	−0.75 **	−0.88 **	−0.91 **

The * means significance of $p < 0.05$, the ** means significance of $p < 0.01$.

As shown in Figure 4, the wind speed range in Chengdu was very small (0.8–1.4 m/s), thus, mass PM concentrations in the Chengdu urban area were high but changed little with wind speed. There was a negative correlation between wind speed and PM_{10}, $PM_{2.5}$ and $PM_{1.0}$, with correlation coefficients of −0.35, −0.56 and −0.62, respectively (Table 1) in the Chengdu urban area. The correlation coefficients of the win speed with atmospheric PM follow the pattern $PM_{1.0} > PM_{2.5} > PM_{10}$. However, the correlation coefficients between wind speed and PM_{10}, $PM_{2.5}$ were not significantly obvious with $p > 0.05$. But the coefficient between wind speed and $PM_{1.0}$ showed clear correlation with $p < 0.05$.

As shown in Figure 5, air pressure values first decreased from January to July, then increased from July to December, showing an inverted parabolic distribution. Mass PM concentrations display a positive relation with air pressure, with significant correlation coefficients ($p < 0.05$) for PM_{10}, $PM_{2.5}$ and $PM_{1.0}$ of 0.71, 0.77 and 0.77, respectively. The surface experiences a convergence up draft when controlled by low pressure; the up draft promotes the dispersion of PM from the ground up into the air, and PM concentrations at the sampling point are subsequently reduced. Conversely, when there was high pressure, the down draft restrains the upward movement of PM, causing an accumulation of particles [40].

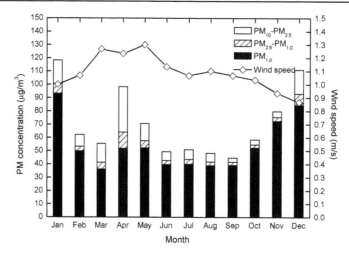

Figure 4. Variations in monthly mass concentrations of PM$_{10}$, PM$_{2.5}$ and PM$_{1.0}$ related to average wind speed.

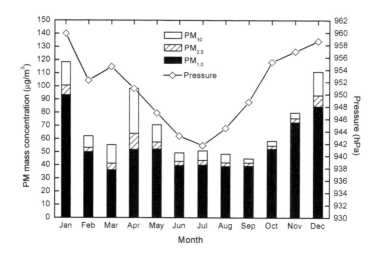

Figure 5. Variations in monthly mass concentrations of PM$_{10}$, PM$_{2.5}$ and PM$_{1.0}$ related to air pressure.

Figure 6 clearly shows that there was a negative relation between the temperature and PM concentrations, with correlation coefficients ($p < 0.01$) of -0.73, -0.75 and -0.75 for PM$_{10}$, PM$_{2.5}$ and PM$_{1.0}$, respectively. High temperatures are clearly conducive to intense convection: Atmospheric PM is transported quickly and effectively, allowing its accelerated dispersion, and thus decreasing local mass concentrations. Conversely, low temperatures and the temperature inversion layer caused by radiative cooling weakens convection [41]; in these circumstances, atmospheric PM remains suspended under the inversion layer, leading to higher atmospheric PM concentrations.

Figure 7 shows that PM concentrations have a negative correlation with precipitation, especially in winter (January and December) and summer (July), with correlation coefficients of -0.59 ($p < 0.05$), -0.58 ($p < 0.05$), and -0.57 ($p > 0.05$), respectively. The influence of precipitation on atmospheric PM is twofold. First, the effect of the microphysical processes of raindrops upon PM, including adsorption and collision, lead to the wet deposition of PM. Second, after rainy weather dust and fugitive dust previously suspended in the atmosphere is notably reduced, leading to a significant decrease in mass PM concentrations [42].

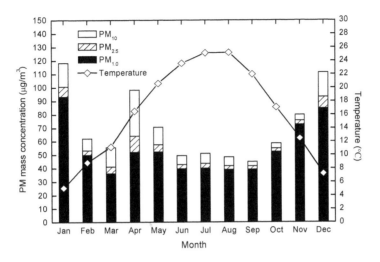

Figure 6. Variations in monthly mass concentrations of PM$_{10}$, PM$_{2.5}$ and PM$_{1.0}$ related to temperature.

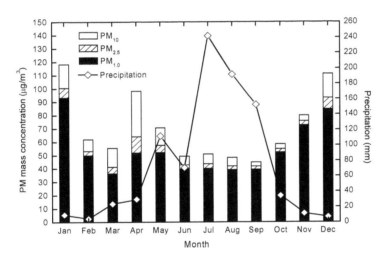

Figure 7. Variations in monthly mass concentrations of PM$_{10}$, PM$_{2.5}$ and PM$_{1.0}$ related to precipitation.

The relation between concentrations of PM$_{10}$, PM$_{2.5}$ and PM$_{1.0}$ and RH are illustrated in Figure 8. It shows that RH in the Chengdu urban area was >70% all year round and that there were higher RH values in fall and summer than that in spring. There was no apparent correlation all year around. The correlation coefficients between RH and PM$_{10}$, PM$_{2.5}$ and PM$_{1.0}$ were less than −0.25. In late summer and early fall, RH values remained high, but because of the high temperatures, unstable atmosphere, intense convection and abundant rainfall, the dispersion of atmospheric PM, wet deposition and cloud condensation nuclei played a leading role. PM pollution, which was caused by the growth of hygroscopicity under high RH conditions, usually in the stable atmospheric strata, but also sometimes when there was a gentle breeze, or within the thermal inversion layer [43].

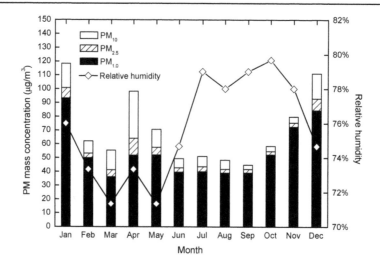

Figure 8. Variations in monthly mass concentrations of PM₁₀, PM₂.₅ and PM₁.₀ related to RH.

3.4. Relation between PM₁₀, PM₂.₅, PM₁.₀ and Horizontal Visibility

There are various ways in which atmospheric visibility can be affected, such as through light scattering and absorption caused by high concentrations of fine particles and the sulfate, nitrate and carbonate matter attached to them [44,45]. In addition, particles experiencing hygroscopic growth affect light extinction through their impact on the atmosphere's particle scattering efficiency. Combining Figure 9 and Table 1, we see that horizontal visibility in the Chengdu urban area is generally low, ranging from 7.5 km in January to 11.0 km in August. There was a negative correlation between horizontal visibility and concentrations of PM₁₀, PM₂.₅ and PM₁.₀, with significant correlation coefficients ($p < 0.01$) of -0.75, -0.88 and -0.91, respectively. When mass concentrations of these three PM sizes exceed 100 µg/m³, visibility is <8.0 km; when concentrations are <40 µg/m³, visibility exceeds 11.0 km.

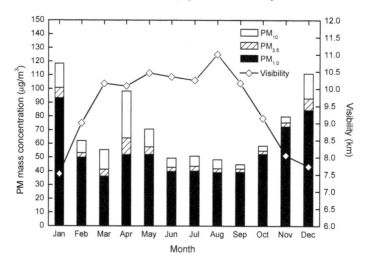

Figure 9. Variations in monthly mass concentrations of PM₁₀, PM₂.₅, and PM₁.₀ related to horizontal visibility.

4. Conclusions

(1) Mass concentrations of PM_{10}, $PM_{2.5}$ and $PM_{1.0}$ were higher in winter and lower in summer from 2009 to 2011 in the Chengdu urban area. Peak average mass concentrations of PM_{10}, $PM_{2.5}$ and $PM_{1.0}$ appeared in January and were 118 $\mu g/m^3$, 101 $\mu g/m^3$ and 93 $\mu g/m^3$ for 2009, 2010 and 2011, respectively. Minimum values occurred in September and were 45 $\mu g/m^3$, 42 $\mu g/m^3$ and 39 $\mu g/m^3$ for 2009–2011, respectively. This shows an improvement in atmospheric PM pollution accompanied by annual falls in PM_{10}, $PM_{2.5}$ and $PM_{1.0}$ concentrations.

(2) Analysis of the correlation between PM_{10}, $PM_{2.5}$ and $PM_{1.0}$ shows that $PM_{10}/PM_{2.5}$, $PM_{10}/PM_{1.0}$ and $PM_{2.5}/PM_{1.0}$ mass concentration correlation coefficients were generally high, reaching 0.92, 0.84 and 0.98, respectively. In addition, $PM_{2.5}$ and $PM_{1.0}$ account for 66% and 54% of PM_{10}, respectively, and $PM_{1.0}$ to $PM_{2.5}$ mass loading reached 89%. This indicates that fine particles are the major contributor to PM air pollution, especially fine particles with an aerodynamic diameter <1 μm.

(3) Except for the relation with air pressure values, significant negative correlations existed between the other meteorological factors (e.g., temperature and precipitation) and PM. Temperature and air pressure influenced the transport and accumulation of PM through convection. Precipitation not only accelerated the wet deposition of PM, but also constricted the movement of surface dust. The correlation between RH and PM concentrations was not obvious. There was an obviously negative correlation between PM and visibility rates. Visibility degradation was primarily attributed to light extinction caused by PM.

This study makes a preliminary analysis of PM pollution and its relation to meteorological factors in Chengdu, representative of the Sichuan Basin urban area, relevant to improving local air quality. The results may support future policies directed at emissions control in China. However, there are still some limitations of the study. There is only one single site PM monitoring in this study because of the instruments limitation. More detailed observations, not only of PM mass concentrations but also of the chemical components and optical properties, from industrial areas, residential areas, and outlying suburbs of the Sichuan Basin are still needed to further study the PM pollution of this region. Also long term observations are needed to study the variation of aerosol in Sichuan Basin. In this study it just analyzed the relationship between the RH and PM basically. Actually, the RH has obvious effect on aerosol particle formation and conversion, more detailed works about the reaction of water vapor on aerosol particles at Sichuan Basin are needed in future.

Acknowledgements

This work is financially supported by grants from the National Key Project of Basic Research (2014CB441201), the Project (41475124 & 41475037) supported by NSFC, CAMS Basis Research Project (2014R17 & 2013Z007).

Author Contributions

Yang Li conceived and designed the experiment. The paper was written by Yang Li with a significant contribution by Quanliang Chen and Hujia Zhao. Lin Wang and Ran Tao analyzed the data and presented the results.

Conflicts of Interest

The authors declare no conflict of interest.

References

1. Charlson, R.J.; Schwartz, S.E.; Hales, J.M.; Cess, D.; Coakley, J.A.; Hansen, J.E. Climate forcing by anthropogenic aerosols. *Science* **1992**, *255*, 423–430.

2. Twomey, S.A.; Piepgrass, M.; Wolfe, T.L. An assessment of the impact of pollution on the global cloud albedo. *Tellus* **1984**, *36*, 356–366.

3. Tao, J.; Chen, J.M.; Chen, X. Impacts of new particle formation on aerosol Cloud Condensation Nuclei (CCN) activity in Shanghai: Case study. *Atmos. Chem. Phys. Discuss.* **2014**, *14*, 18641–18677.

4. Sloane, C.S.; Watson, J.; Chow, J.; Pritchett, L.; Willard, R.L. Size-segregated fine particle measurements by chemical species and their impact on visibility impairment in Denver. *Atmos. Environ.* **1991**, *25*, 1013–1024.

5. Chang, D.; Song, Y.; Liu, B. Visibility trends in six megacities in China 1973–2007. *Atmos. Res.* **2009**, *94*, 161–167.

6. Che, H.Z.; Zhang, X.Y.; Li, Y.; Zhou, Z.J.; Qu, J.J.; Hao, X.J. Haze trends over the capital cities of 31 provinces in China, 1981–2005. *Theor. Appl. Climatol.* **2009**, *97*, 235–242.

7. Leng, C.P.; Cheng, T.T.; Zhang, R.J.; Tao, J.; Chen, J.M.; Zha, S.P.; Zhang, Y.W.; Zhang, D.Q.; Li, X.; Kong, L.D. Variation of Cloud Condensation Nuclei (CCN) and aerosol activity during fog-haze episode: A case study from Shanghai. *Atmos. Chem. Phys. Discuss.* **2014**, *14*, 16997–17036.

8. Du, H.H.; Kong, L.D.; Cheng, T.T.; Li, L.; Du, J.F.; Chen, J.M.; Xia, X.A.; Leng, C.P.; Huang, G.H. Insights into summertime haze pollution events over Shanghai based on online water-soluble ionic composition of aerosols. *Atmos. Environ.* **2011**, *45*, 5131–5137.

9. Hong, Y.C.; Lee, J.T.; Kim, H.; Ha, E.H.; Schwartz, J.; Christiani, D.C. Effects of air pollutants on acute stroke mortality. *Environ. Health Persp.* **2002**, *110*, 187–191.

10. Zhang, R.; Jing, J.; Tao, J.; Hsu, S.C.; Wang, G.; Cao, J.; Lee, C.S.L.; Zhu, L.; Chen, Z.; Zhao, Y.; Shen, Z. Chemical characterization and source apportionment of $PM_{2.5}$ in Beijing: Seasonal perspective. *Atmos. Chem. Phys.* **2013**, *13*, 7053–7074.

11. Zhang, R.J.; Tao, J.; Ho, K.F.; Shen, Z.X.; Wang, G.H.; Cao, J.J.; Liu, S.X.; Zhang, L.M.; Lee, S.C. Characterization of atmospheric organic and elemental carbon of $PM_{2.5}$ in a typical semi-arid area of Northeastern China. *Aerosol Air Qual.* Res. **2012**, *12*, 792–802.

12. Zhao, H.J.; Che, H.Z.; Zhang, X.Y. Characteristics of visibility and Particulate Matter (PM) in an urban area of Northeast China. *Atmos. Pollut. Res.* **2013**, *40*, 427–434.

13. Cheng, T.T.; Wang, H.; Xu, Y.F.; Li, H.Y. Climatology of aerosol optical properties in Northern China. *Atmos. Environ.* **2006**, *40*, 1495–1509.

14. Cheng, T.T.; Zhang, R.J.; Han, Z.W.; Fang, W. Relationship between ground-based particle component and column aerosol optical property in dusty days over Beijing. *Geophys. Res. Lett.* **2008**, doi:10.1029/2008GL035284.

15. Wang, X.; Huang, J.; Zhang, R.; Chen, B.; Bi, J. Surface measurements of aerosol properties over Northwest China during ARM China 2008 deployment. *J. Geophys. Res.* **2010**, doi:10.1029/2009JD013467.

16. Jia, X.; Cheng, T.T.; Chen, J.M.; Xu, J.W.; Chen, Y.H. Columnar optical depth and vertical distribution of aerosols over Shanghai. *Aerosol Air Qual. Res.* **2012**, *12*, 316–326.

17. Xu, J.W.; Tao, J.; Zhang, R.J.; Cheng, T.T.; Leng, C.P.; Chen, J.M.; Huang, G.H.; Li, X.; Zhu, Z.Q. Measurements of surface aerosol optical properties in winter of Shanghai. *Atmos. Res.* **2012**, *109–110*, 25–35.

18. Cao, J.J.; Lee, S.C.; Ho, K.F.; Zhang, X.Y.; Zou, S.C.; Fung, K.; Chow, J.C.; Watson, J.G. Characteristics of carbonaceous aerosol in Pearl River Delta Region, China during 2001 winter period. *Atmos. Environ.* **2003**, *37*, 1451–1460.

19. GB3095–2012 Ambient Air Quality Standards. Available online: http://kjs.mep.gov.cn/hjbhbz/bzwb/dqhjbh/dqhjzlbz/201203/t20120302_224165.htm (accessed on 10 November 2014). (In Chinese)

20. Luo, Y.F.; Li, W.L.; Zhou, X.J.; He, Q.; Qing, J.Z. Analysis of the atmospheric aerosol optical depth over China in 1980s. *Acta Meteorol. Sini.* **2000**, *14*, 490–502.

21. Li, C.C.; Mao, J.T.; Lau, K.H. Characteristics of the aerosol optical depth distributions over Sichuan Basin derived from MODIS data. *J. Appl. Meteorol.* **2003**, *14*, 1–17. (In Chinese)

22. Sichuan Environment Condition Communique of 2009. Available online: http://www.schj.gov.cn/cs/hjjc/zkgg/ (accessed on 10 November 2014).

23. Sichuan Environment Condition Communique of 2010. Available online: http://www.schj.gov.cn/cs/hjjc/zkgg/ (accessed on 10 November 2014).

24. Sichuan Environment Condition Communique of 2011. Available online: http://www.schj.gov.cn/cs/hjjc/zkgg/ (accessed on 10 November 2014).

25. Tao, J.; Zhang, L.M.; Engling, G.; Zhang, R.J.; Yang, Y.H.; Cao, J.J.; Zhu, C.S.; Wang, Q.Y.; Luo, L. Chemical composition of PM2.5 in an urban environment in Chengdu, China: Importance of springtime dust storms and biomass burning. *Atmos. Res.* **2013**, *122*, 270–283.

26. Tao, J.; Gao, J.; Zhang, L.; Zhang, R.; Che, H.; Zhang, Z.; Lin, Z.; Jing, J.; Cao, J.; Hsu, S.-C. PM2.5 pollution in a megacity of southwest China: Source apportionment and implication. *Atmos. Chem. Phys.* **2014**, *14*, 8679–8699.

27. Tao, R.; Che, H.Z.; Chen, Q.L.; Tao, J.; Wang, Y.Q.; Sun, J.Y.; Wang, H.; Zhang, X.Y. Study of aerosol optical properties based on ground measurements over Sichuan Basin, China. *Aerosol Air Qual. Res.* **2014**, *14*, 905–915.

28. Bulletin on the State of the Environment in Chengdu in 2009. Available online: http://www.cdepb.gov.cn/upload/ (accessed on 10 November 2014).

29. Bulletin on the State of the Environment in Chengdu in 2010. Available online: http://www.cdepb.gov.cn/upload/ (accessed on 10 November 2014).

30. Bulletin on the State of the Environment in Chengdu in 2011. Available online: http://www.cdepb.gov.cn/upload/ (accessed on 10 November 2014).

31. Bulletin on the State of the Environment in Chengdu in 2012. Available online: http://www.cdepb.gov.cn/upload/ (accessed on 10 November 2014).

32. Tao, J.; Cheng, T.T.; Zhang, R.J.; Cao, J.J.; Zhu, L.H.; Wang, Q.Y.; Luo, L.; Zhang, L.M. Chemical composition of PM$_{2.5}$ at an urban site of Chengdu in Southwestern China. *Adv. Atmos. Sci.* **2013**, *30*, 1070–1084.

33. Chen, B.H.; Li, Y.Q. Relationships between air particle and meteorological elements in Chengdu. *Urban Environ. Urban. Ecol.* **2006**, *19*, 18–24.

34. Hu, M.; Liu, S.; Wu, Z.J. Effects of high temperature, high relative humidity and rain process on particle size distributions in the summer of Beijing. *Environ. Sci.* **2006**, *27*, 2293–2298.

35. Yuan, Y.; Liu, D.; Che, R. Research on the pollution situation of atmospheric particulates during autumn in Beijing City. *Ecol. Environ.* **2007**, *16*, 18–25.

36. Deng, L.Q.; Qian, J.; Liao, R.X. Pollution characteristics of atmospheric particulates in Chengdu from August to September in 2009 and their relationship with meteorological conditions. *China Environ. Sci.* **2012**, *32*, 1433–1438. (In Chinese)

37. Yao, Q.; Cai, Z.Y.; Zhang, C.C. Variety characteristics and influence factors of aerosol mass concentrations in Tianjin City. *Ecol. Environ. Sci.* **2010**, *19*, 2225–2231.

38. Galindo, N.; Varea, M.; Gil-Moltó, J.; Yubero, E.; Nicolás, J. The influence of meteorology on particulate matter concentrations at an urban mediterranean location. *Water Air Soil Pollut.* **2011**, *215*, 365–372.

39. Pateraki, S.; Asimakopoulos, D.N.; Flocas, H.A.; Maggos, T.; Vasilakos, C. The role of meteorology on different sized aerosol fractions (PM$_{10}$, PM$_{2.5}$, PM$_{2.5-10}$). *Sci. Total Environ.* **2012**, *419*, 124–135.

40. Ren, Z.H.; Su, F.Q.; Chen, Z.H. Influence of synoptic systems on the distribution and evolution process of PM$_{10}$ concentration in the boundary layer in summer and autumn. *Chin. J. Atmos. Sci.* **2008**, *32*, 741–751

41. Lin, J.; Liu, W.; Yan, I. Relationship between meteorological conditions and particle size distribution of atmospheric aerosols. *J. Meteorol. Environ.* **2009**, *25*, 1–5.

42. Shen, Z.Y.; Lu, B.; Chen, H.B. Effects of meteorology condition in precipitation on aerosol concentration in Zhengzhou. *Meteorol. Environ. Sci.* **2009**, *32*, 55–58.

43. Mu, C.Y.; Tu, Y.Q.; Feng, Y. Effect analysis of meteorological factors on the inhalable particle matter concentration of atmosphere in Hami. *Meteorol. Environ. Sci.* **2011**, *34*, 75–79.

44. Lin, J.J. Characterization of water-soluble ion species in urban ambient particles. *Environ. Int.* **2002**, *28*, 55–61.

45. Qu, W.J.; Wang, J.; Zhang, X.Y.; Wang, D.; Sheng, L.F. Influence of relative humidity on aerosol composition: Impacts on light extinction and visibility impairment at two sites in coastal area of China. *Atmos. Res.* **2015**, *153*, 500–511.

5

Characteristics of Black Carbon Aerosol during the Chinese Lunar Year and Weekdays in Xi'an, China

Qiyuan Wang [1], **Suixin Liu** [1], **Yaqing Zhou** [1], **Junji Cao** [1,2,*], **Yongming Han** [1], **Haiyan Ni** [1], **Ningning Zhang** [1] **and Rujin Huang** [1,3,4]

[1] Key Laboratory of Aerosol Chemistry & Physics, Institute of Earth Environment, Chinese Academy of Sciences, Xi'an 710061, China; E-Mails: wangqy@ieecas.cn (Q.W.); lsx@ieecas.cn (S.L.); zhouyq@ieecas.cn (Y.Z.); yongming@ieecas.cn (Y.H.); nihy@ieecas.cn (H.N.); zhangnn@ieecas.cn (N.Z.); Rujin.Huang@psi.ch (R.H.)

[2] Institute of Global Environmental Change, Xi'an Jiaotong University, Xi'an 710049, China

[3] Lab of Atmospheric Chemistry, Paul Scherrer Institute (PSI), Villigen 5232, Switzland

[4] Centre for Climate and Air Pollution Studies, Ryan Institute, National University of Ireland Galway, University Road, Galway, Ireland

* Author to whom correspondence should be addressed; E-Mail: cao@loess.llqg.ac.cn

Academic Editor: Robert W. Talbot

Abstract: Black carbon (BC) aerosol plays an important role in climate forcing. The net radiative effect is strongly dependent on the physical properties of BC particles. A single particle soot photometer and a carbon monoxide analyser were deployed during the Chinese Lunar Year (CLY) and on weekdays at Xi'an, China, to investigate the characteristics of refractory black carbon aerosol (rBC). The rBC mass on weekdays ($8.4 \ \mu g \cdot m^{-3}$) exceeds that during the CLY ($1.9 \ \mu g \cdot m^{-3}$), presumably due to the lower anthropogenic emissions during the latter. The mass size distribution of rBC shows a primary mode peak at ~205 nm and a small secondary mode peak at ~102-nm volume-equivalent diameter assuming $2 \ g \cdot cm^{-3}$ in void-free density in both sets of samples. More than half of the rBC cores are thickly coated during the CLY ($f_{BC} = 57.5\%$); the percentage is slightly lower ($f_{BC} = 48.3\%$) on weekdays. Diurnal patterns in rBC mass and mixing state differ for the two sampling periods, which are attributed to the distinct anthropogenic activities. The rBC mass and CO mixing ratios are strongly correlated with slopes of 0.0070 and 0.0016 $\mu g \cdot m^{-3} \cdot ppbv^{-1}$ for weekdays and the CLY, respectively.

Keywords: refractory black carbon; size distribution; mixing state; carbon monoxide

1. Introduction

Black carbon (BC) aerosol, a by-product of the incomplete combustion of fossil fuels and biomass, plays a unique and important role in Earth's climate system. In the boundary layer, these particles absorb solar radiation and cool the Earth's surface, but at the top of the atmosphere, they add to the climate warming caused by radiatively-active gases [1,2]. The total climate forcing of BC aerosol has been estimated to be $+1.1$ W·m^{-2}, which is ranked as the second largest contributor to anthropogenic radiative forcing after CO_2 [3]. Coated BC particles also can act as cloud condensation nuclei (CCN), thereby contributing to the indirect forcing of climate [4]. In addition to the effects on the environment, epidemiological research suggests that BC particles have adverse health effects: they have been implicated in respiratory and cardiovascular diseases [5,6]. Once emitted, the BC aerosol can be transported over regional to intercontinental scales, and the particles are ultimately removed from the atmosphere through wet (e.g., in precipitation) and dry deposition to the Earth's surface. The resulting average atmospheric lifetime for BC is about a week [3]. Therefore, reductions in BC aerosol emissions are an attractive option, not only for mitigating global warming, but also for improving environmental conditions in other ways [7].

Most large Chinese cities have suffered from air pollution in recent years (e.g., [8–10]), and this has become one of the nation's top environmental concerns. An anthropogenic emission inventory of China developed from 1990 to 2005 shows that the BC emissions, and, therefore, presumably, concentrations, have increased since 2000 [11]. Most studies (e.g., [12,13]) in China have used filter-based techniques that measure bulk aerosol absorption rather than the BC mass concentrations directly. Previous studies indicate that key properties of the BC aerosol, that is the particle's concentration, size, shape and mixing state, vary in complex ways and that these properties depend on many interacting environmental factors [14,15]. Due to the systematic limitations of most of the current filter-based BC measurements [3], direct examination of the BC size distribution and mixing state are not possible. Therefore, high-quality measurements of BC emissions have become increasingly important for accurately assessing the BC impacts on the atmospheric radiative energy balance, human health and air quality.

Here, we report the results of a study on individual refractory black carbon (rBC) mass concentrations, size distributions, the mixing state and the relationship with carbon monoxide (CO). The terms, rBC and BC, are operationally defined based on the applied measurement technique (laser-induced incandescence and light absorption), and they both refer to the most light absorbing component of carbonaceous combustion particles [16,17]. Samples for the study were collected from Xi'an, China, during the Chinese Lunar Year (CLY), and we contrast those data with observations taken on weekdays. The CLY is one of the most important traditional festivals in China. Numerous migrant workers return to their hometowns to reunite with their families and to celebrate the festival. During this one-week holiday, most companies and many industries shut down, and traffic is also at a minimum. The concentration of PM$_{2.5}$ (particulate matter with an aerodynamic diameter of less than 2.5 micrometres) during this period (128.9 µg·m^{-3}) is much lower than on weekdays (191.4 µg·m^{-3}).

This gives researchers a once-a-year opportunity to study the atmosphere when anthropogenic emissions are greatly reduced. In contrast, normal weekdays (Monday to Friday) are arguably representative of a more typical scenario with respect to pollution emissions. The comparisons of aerosol properties made for the holiday *versus* weekdays will lead to a better understanding of the impact of anthropogenic emissions on the physical properties of the rBC aerosol in China.

2. Experimental Section

2.1. Research Site

Xi'an, one of the major tourist cities in China, is located on the Guanzhong Plain at 34°27′N, 108°95′E, and it has a population of >8 million. Measurements of rBC were made from the rooftop (~10 m above ground level) of the Institute of Earth Environment, Chinese Academy of Sciences (IEECAS; [18,19]). This monitoring site is located in an urban zone surrounded by a residential area ~15 km south of downtown Xi'an, where there are no major industrial activities or local fugitive dust sources (see Figure 1). The sampling periods were from 9 to 15 February 2013 on the Chinese Lunar Year (CLY) and from 23 December 2012 to 18 January 2013 on weekdays. The results of New Year's Eve (9 to 10 February) are not included in the discussion in the following section to exclude the impacts of the intensive fireworks displays. Our focus is on the difference in the results due to the reduced anthropogenic emissions during the holiday.

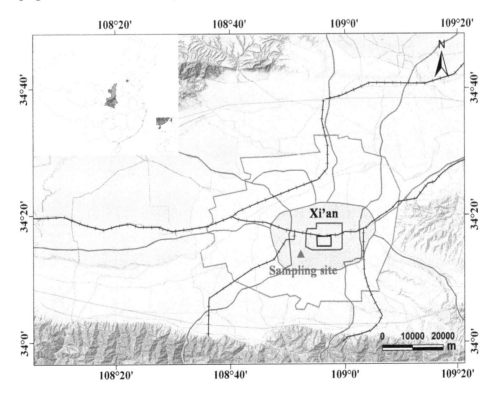

Figure 1. Location of the Xi'an sampling site.

2.2. rBC and CO Measurements

The commercially available single particle soot photometer (SP2) instrument (Droplet Measurement Technology, Boulder, CO, USA) has been proven useful for measuring rBC mass, size and mixing state [20–22]. The operating principles of the SP2 have been described in detail previously [23,24]. Briefly, the SP2 relies on laser-induced incandescence to quantify the rBC mass of individual particles. Continuous intracavity Nd:YAG laser light at 1064 nm is used to heat BC-containing particles to their vaporization point. The peak incandescence signal is linearly related to the rBC mass in the particle, irrespective of the particle morphology or mixing state; this holds true over most of the rBC mass range typically observed in the accumulation mode [25].

Here, the rBC mass in the range ~0.4–1000 fg, equivalent to ~70–1000 nm volume equivalent diameter (VED) is quantified, assuming a void-free density of 2.0 g·cm^{-3}. This range covers >90% of the rBC mass in the accumulation mode. Note that the SP2 only quantifies the mass of the most refractory and most efficient light-absorbing component of combustion aerosol. We use the term "BC-containing particle" to refer to any particle containing this material, whether or not non-rBC material is present.

The incandescence signal was used to obtain the single particle rBC masses after calibration using a fullerene soot standard reference sample (Lot F12S011, AlphaAesar, Inc., Ward Hill, Massachusetts). The fullerene soot was separated by size by installing a differential mobility analyser (EPS-20 Electrical Particle Size, HCT Co. Ltd., Icheon-si, Korea) upstream of the SP2. Calibration curves were obtained from the peak intensities of the incandescence signals for the fullerene soot particles over a range of masses corresponding to rBC of ~0.35–30 fg; this was based on the mass-mobility relationships for this material described in Moteki and Kondo [26]. This mass range corresponds to ~80–450 nm VED, over which the calibration was close to purely linear: the various determinations of the mass to mobility relationship for this material are in good agreement [16,26].

The total uncertainty in the rBC mass determinations is ~25%; this is estimated by propagating the uncertainties in: (1) the rBC mass calibrations, including possible variability in the SP2's response to ambient rBC mass (~20%, [26,27]); (2) sample flow measurements (~10%); and (3) estimates of the rBC mass outside of the SP2 detection range (~10%). The concentrations of rBC particles are corrected to standard temperature and pressure (STP: 273.15 K, 1013.25 hPa).

The SP2 has the capability of determining the rBC mixing state. The delay time between the peaks from the scattered light and incandescence signals has been used as an indicator of the amount of non-rBC material internally mixed with individual rBC particles [22,23,28]. This is a commonly-used approach for distinguishing "thinly-" from "thickly-coated" rBC particles, and it is sensitive to optically significant amounts of non-rBC material. The delay occurs because the coatings must be removed from the rBC fractions before the onset of incandescence. The number fraction of thickly-coated rBC (f_{BC}) based on the distribution of delay times measured by the SP2 is an indication of the degree to which the rBC particles are coated with other substances. High values for f_{BC} are consistent with more aged rBC particles, and this is due to the various processes in the ambient atmosphere that lead to the formation of coatings. In this study, the time criterion for distinguishing uncoated or thinly-coated rBC and thickly-coated rBC is 2 μs, and this is based on the observed minimum in the bimodal frequency distribution of delay times [29]. The SP2 measurement data are processed for a 1-h average for the later data analysis and discussion.

Five-minute average mixing ratios of carbon monoxide (CO) are obtained using gas-filter correlation technology with infrared photometric detection. The carbon monoxide analyser (Model

EC9830T, Ecotech Pty Ltd., Knoxfield, Australia) used for these analyses has a detection limit for the CO of 20 ppbv.

2.3. Local Meteorological Conditions

Hourly temperature, relative humidity (RH) and wind speed were measured using an automatic weather station (MAWS201, Vaisala, Vantaa, Finland) configured with an RH/temperature probe (Model QMH101) and wind sensor (Model QMW101-M2). The meteorological conditions during the campaign are presented in Figure 2. The temperature has obvious diurnal cycles, with higher values observed during the daytime. Relative humidity shows peak values during the night. The wind speed is relatively low most of the time. For CLY samples, the average temperature, RH and wind speed are 4.7 ± 3.4 °C, $57\% \pm 23\%$ and 0.2 ± 0.2 m·s^{-1}, respectively. In comparison, the average temperature, RH and wind speed are 0.6 ± 3.6 °C, $38\% \pm 13\%$ and 0.3 ± 0.3·m·s^{-1} on weekdays, respectively.

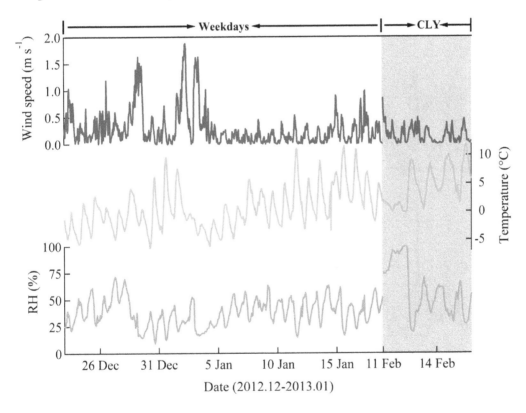

Figure 2. Time series of hourly averaged relative humidity (RH), temperature and wind speed during the Chinese Lunar Year (CLY) and on weekdays.

3. Results and Discussion

3.1. rBC Loadings

The hourly average rBC number and mass concentrations, as well as f_{BC} values for the CLY and weekday samples are shown in Figure 3, and a statistical summary of the data is presented in Table 1. For the CLY samples, the number and mass concentrations of rBC range from 99 to 1231 cm^{-3} and 0.2 to 3.9 μg·m^{-3}, with an average value of 503 ± 212 cm^{-3} and 1.9 ± 0.8 μg·m^{-3}, respectively. The concentration of rBC during CLY is similar to that during clean periods, but much lower than polluted periods in Xi'an [22]. The average rBC mass concentration in this study is comparable to or lower than the values during the 2010 Shanghai World Expo (2.0 $\mu g \cdot m^{-3}$, measured with an SP2, [30]) and the 2008 Beijing Olympic Games (3.2 $\mu g \cdot m^{-3}$, measured with an Aethalometer, [31]). Anthropogenic emissions are reduced during these events in Shanghai and Beijing, because the government implements a series of air pollution control measures to improve the air quality; these include reducing emissions from coal-fired power plants and strict regulations on motor vehicle usage. In comparison, the rBC number and mass concentrations during weekdays vary from 65 to 6505 cm^{-3} and 0.2 to 38.8 $\mu g \cdot m^{-3}$, with an average value of $1{,}850 \pm 1242$ cm^{-3} and 8.4 ± 6.5 $\mu g \cdot m^{-3}$, respectively. The high standard deviation suggests that there is a large BC burden on weekdays. The average rBC value on weekdays exceeds the average value of the CLY by a factor of approximately five. The enhanced rBC loadings on weekdays are almost certainly due to the greater impacts from the ongoing anthropogenic activities that pollute the air.

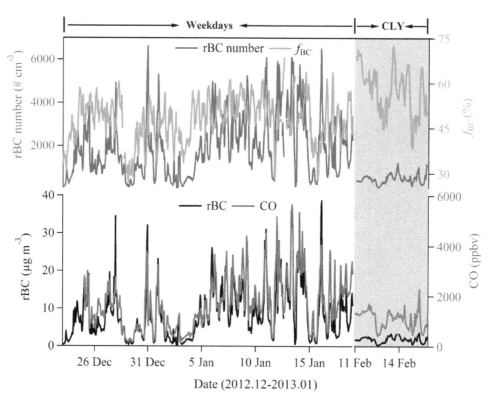

Figure 3. Time series of the hourly averaged refractory black carbon (rBC) number and mass concentrations, the number fraction of thickly-coated rBC and the CO mixing ratio during the Chinese Lunar Year (CLY) and on weekdays.

Table 1. Summary of the rBC number and mass concentrations, as well as the number fractions of thickly-coated rBC (f_{BC}) during the Chinese Lunar Year and on weekdays.

	Chinese Lunar Year			Weekdays		
	rBC (cm^{-3})	rBC ($\mu g \cdot m^{-3}$)	f_{BC} (%)	rBC (cm^{-3})	rBC ($\mu g \cdot m^{-3}$)	f_{BC} (%)
Average	503	1.9	57.5	1850	8.4	48.3
SD	212	0.8	8.2	1242	6.5	6.9
Maximum	1231	3.9	76.1	6505	38.8	63.6
Minimum	99	0.2	26.7	65	0.2	27.3

In Figure 4, the diurnal variations in rBC mass, as well as f_{BC} measured over Xi'an during the CLY and weekdays are plotted. A clear diurnal variation in rBC concentrations is observed for the weekdays with a broad nocturnal maximum from around 22:00 to 03:00 local standard time (LST), which is consistent with rBC variation during the polluted period in Xi'an [22]. This maximum can be explained by the development of a shallow and stable boundary layer caused by nocturnal radiative cooling; this leads to the build-up of pollutants, including emissions from evening rush-hour traffic, residential heating during cold weather and diesel vehicles at midnight. The rBC loadings then are maintained at a relatively high level. This stability condition is followed by a sharp increase to the daytime peak concentrations from 08:00 to 09:00 LST, which can be attributed to the morning rush-hour traffic. The increased solar heating as the day advances produces a deeper and more turbulent boundary layer, and that leads to greater dispersion and dilution of rBC in the near surface air. Finally, the rBC concentrations decrease gradually, and a diurnal minimum is reached in the afternoon (14:00–17:00 LST). The diurnal variations in rBC mass during the CLY show different features from the weekdays, and those differences are likely linked to local dynamics. The maximum rBC mass loadings occur between 09:00 and 10:00 LST, and the high loadings are followed by a sharp decrease to daytime lows from 14:00 to 17:00 LST. The mass increases again to reach a nocturnal peak around 20:00–21:00 LST.

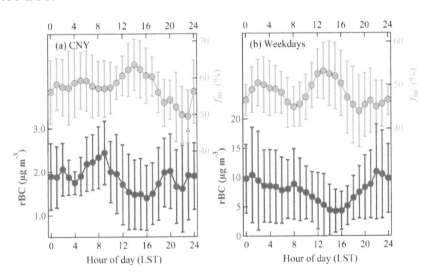

Figure 4. Diurnal variations of hourly averaged rBC mass (red dots) and the number fraction of thickly-coated rBC particles (f_{BC}) (green dots) during (**a**) the Chinese Lunar Year and (**b**) on weekdays. Vertical lines indicate one standard deviation. LST stands for local standard time.

3.2. rBC Mass Size and Mixing State

The mixing state of BC aerosol is one of the major uncertainties in the models used to evaluate BC direct radiative forcing [3]. As shown in Table 1, the hourly average f_{BC} ranges from 26.7% to 76.1%, with an average of 57.5% during the CLY; this high percentage of thickly-coated particles shows that rBC aerosol is mainly composed of aged particles during the festival. The remaining particles are classified as either "thinly-coated" or "uncoated". In comparison, on weekdays, f_{BC} varies from 27.3%

to 63.6%, with an average of 48.3%, which is ~20% lower than during the CLY. Due to the high loadings of particulates on weekdays, the organics may be the primary contributor to rBC coatings [22].

The diurnal patterns in f_{BC} during the CLY and weekdays show similar trends, with high values in the afternoon and low values at night (Figure 4), which is consistent with the f_{BC} pattern during the polluted period in Xi'an [22]. Generally, the afternoon f_{BC} peaks in both sets of samples are anti-correlated with the rBC concentrations. There are two possible mechanisms to explain this relationship. First, greater convection in the afternoon would bring more aged air masses from the upper atmosphere to the sampling site, and this aged air would tend to contain more internally-mixed rBC particles [14,15]. Second, the strong solar irradiance in the mid-afternoon would favour photochemical oxidation processes that would lead to the mixing of rBC with other chemical species. On weekdays, the low f_{BC} in the early morning and late evening can be attributed to increases in fresh, externally-mixed, rBC particles from vehicles during rush hour.

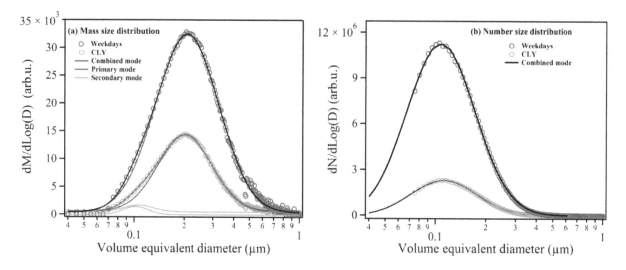

Figure 5. (a) Mass and (b) number size distributions of rBC for the Chinese Lunar Year (CLY) and on weekdays. A two-mode lognormal function fits the data of mass size distribution, while a single-mode lognormal function fits the data of number size distribution. "M" and "D" in the vertical labels represent rBC mass and diameter, respectively.

Figure 5 compares the rBC number and mass size distributions for the CLY and weekday samples. It should be noted that the rBC size as used here refers to the VED of the "core", and that does not include the contribution of the non-rBC material to the particle diameter. It is found that the number size distributions for CLY and weekday rBC particles show a single lognormal distribution, but a two-mode lognormal function is a better fit for the mass size distribution. Furthermore, the mass size distributions during the CLY and weekdays exhibit similar patterns. The primary mode has a peak at ~205 nm, consistent with the results obtained in the lower troposphere at a site in Texas, USA [32], the surface atmosphere of San Jose, USA [14], and Shenzhen, China [28]. A secondary mode in our samples peaks at ~102 nm, and it contains ~5% of the total rBC mass. Wang *et al.* [33] and Huang *et al.* [34] find a larger secondary mode in rural areas of Qinghai Lake (mass median diameter, MMD: 495 nm) and Kaiping (MMD: 690 nm), China, where biofuel/biomass burning are the main sources for rBC. Previous studies have shown that rBC particles from biomass burning tend to be rather large [35], while motor vehicles generally emit smaller rBC particles [36,37]. Thus, the small secondary size mode in Xi'an may be the result of vehicular emissions. However, further studies of rBC emission sources should be conducted to confirm this possibility.

2.3. Relationship between rBC and CO

BC has been found to be correlated with CO in urban areas [12,38], and this is reasonable, because both species are produced by the incomplete combustion of carbon-based fuels. The BC and CO relationship is controlled by the balance between emission sources and sinks. The emission ratios of BC/CO also can vary significantly with the types of emission sources. For example, the BC/CO emission ratios are known to be much higher for diesel engines compared with those for gasoline engines [39]. Thus, the BC/CO ratios can indicate the relative contributions of different sources. In addition, region-specific BC/CO values can be used to evaluate the regional emission inventory of BC and CO [40], and it is also an important constraint on global and regional air quality models [41].

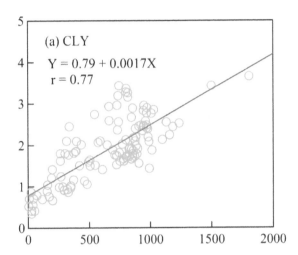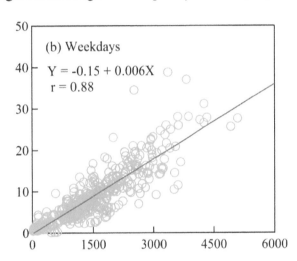

Figure 6. Scatter plots of hourly rBC concentrations *versus* ΔCO mixing ratios during the (**a**) Chinese Lunar Year (CLY) and (**b**) weekdays. The solid lines show linear fits to the data.

Furthermore, the relationship between BC and CO typically changes with time; this is because the atmospheric lifetime of BC is much shorter than that of CO, primarily due to the fact that BC is removed by wet deposition, but CO is not. As a result, the rBC/CO ratios are most informative after correcting for the rBC and CO background. The background concentrations of CO for the CLY and weekdays are defined as the 1.25th percentile of the CO values [38], while the background of the rBC concentration is assumed as zero [42]. The rBC concentrations observed during the CLY and weekdays are plotted against the background-corrected CO concentrations (ΔCO) in Figure 6. A relatively strong relationship between rBC and CO is observed for the weekdays (correlation coefficient, $r = 0.88$), indicating that the two substances are probably controlled by similar sources. Considering the large quantities of coal burned in winter and the heavy workday traffic, the slope of $0.0060\ \mu g \cdot m^{-3} \cdot ppbv^{-1}$ presumably represents the rBC/ΔCO ratio typical of these emissions in winter. The rBC also correlates well ($r = 0.77$) with ΔCO during the CLY with a slope of $0.0017\ \mu g \cdot m^{-3} \cdot ppbv^{-1}$. Compared to weekdays, the rBC/ΔCO ratio is much lower during the CLY.

This may be attributed to the reduced anthropogenic activities during the CLY, such as coal burning and diesel vehicles.

For further perspective, the rBC/ΔCO ratio derived from this study is compared with the results of other studies (Table 2). The rBC/ΔCO ratios during two sampling periods here are within the range (0.0008–0.0062 μg·m^{-3}·ppbv^{-1}) measured in the boundary layer over Europe [43]. The rBC/ΔCO ratio for the CLY is lower than the values for most of the cities listed in Table 2. The rBC/ΔCO ratio on weekdays is towards the upper limit of the values in Beijing (0.0035 to 0.0058 μg·m^{-3}·ppbv^{-1}; [12]). Kondo *et al.* [40] reports values of 0.0057 and 0.0063 μg·m^{-3}·ppbv^{-1} for Tokyo and Nagoya, Japan; these are both comparable to what we observed on weekdays in Xi'an. In contrast, the rBC/CO slope on weekdays in Xi'an is higher than that of Mexico City (0.001 μg·m^{-3}·ppbv^{-1}; [44]) or Guangzhou, China (0.0045 μg·m^{-3}·ppbv^{-1}; [45]).

Table 2. Comparison of rBC concentrations to background-corrected CO mixing ratios (rBC/ΔCO) in Xi'an with other areas.

Location	Sampling Period	rBC/ΔCO	Method [a]	Reference
Xi'an, China	December 2012/January 2013	0.0060	SP2	This study
	February 2013	0.0017		
Beijing, China	November 2005–October 2006	0.0035–0.0058	TOT	[12]
Guangzhou, China	July 2006	0.0045	EC-OC aerosol analyser	[45]
Multiple sites in Europe	April/May/September 2008	0.0008–0.0062	SP2	[44]
Tokyo, Japan	May 2003–February 2005	0.0057	TOT	[40]
Nagoya, Japan	March 2003	0.0063	Light absorption	[40]
Mexico	April 2003/2005	0.001	SP2	[45]

Notes: [a] SP2 stands for single particle soot photometer; TOT is thermal optical transmittance; EC is elemental carbon and OC is organic carbon.

4. Conclusions

An SP2 and a CO instrument were deployed at Xi'an to: (1) characterize the rBC mass, size distribution and mixing state; and (2) investigate the relationship between rBC and CO. Comparisons were made between samples collected during the Chinese Lunar Year (CLY), when anthropogenic emissions were reduced, and samples collected on normal weekdays. The average rBC mass concentration on weekdays (8.4 μg·m^{-3}) exceeds the average value for the CLY (1.9 μg·m^{-3}) by almost five-fold. Different diurnal variations in rBC are observed for the CLY and weekday samples, and this can be explained by the pollution prevention measures during the holiday. The average number fraction of thickly-coated rBC (f_{BC}) during the CLY is 57.5%, suggesting that aged particles compose the major fraction of the rBC aerosol. In comparison, the average f_{BC} is considerably lower (48.3%) on weekdays. Higher f_{BC} values for both sampling periods are observed in the afternoon; this is likely due to the formation of photochemical oxidation products and stronger influences from vertical transport during the afternoon. The rBC size distributions for both sampling periods are lognormal and bi-modal; both have a primary peak at ~205 nm, which most likely reflects the variety of rBC sources in Xi'an. The peak diameter of the secondary mode is ~102 nm, and this secondary peak represents only ~5% of the total rBC mass. This persistent secondary mode may be associated with motor vehicle emissions. Strong correlations are observed between rBC and CO during the CLY celebration and on weekdays, indicating that the two species are controlled by similar sources. The average rBC/ΔCO ratio on weekdays is 0.0060 μg·m^{-3}·ppbv^{-1}, and this is arguably representative of air masses impacted by the pervasive pollution emissions. However, reduced anthropogenic activities during the CLY lead to a decrease in the rBC/ΔCO ratio (0.0017 μg·m^{-3}·ppbv^{-1}) compared with weekdays.

Acknowledgments

This study is supported by the National Natural Science Foundation of China (41230641, 41271481, 40925009) and projects from the "Strategic Priority Research Program" of the Chinese Academy of Science (XDB05060500, XDA05100401), the Ministry of Science & Technology (2012BAH31B03) and the Shaanxi Government (2012KTZB03-01-01).

Author Contributions

This work has been performed in collaboration among all of the authors. Junji Cao, Yongming Han, Rujin Huang and Qiyuan Wang conceived of and designed the experiments. Qiyuan Wang, Suixin Liu, Yaqing Zhou, Haiyan Ni and Ningning Zhang performed the experiments. Qiyuan Wang carried out the main part of the writing, although all co-authors have contributed to the text.

Conflicts of Interest

The authors declare no conflict of interest.

References

1. Ramanathan, V.; Carmichael, G. Global and regional climate changes due to black carbon. *Nat. Geosci.* **2008**, *1*, 221–227.
2. Chung, C.E.; Lee, K.; Mueller, D. Effect of internal mixture on black carbon radiative forcing. *Tellus* **2012**, *64*, 1–13.
3. Bond, T.C.; Doherty, S.J.; Fahey, D.W.; Forster, P.M.; Berntsen, T.; de Angelo, B.J.; Flanner, M.G.; Ghan, S.; Kärcher, B.; Koch, D.; *et al.* Bounding the role of black carbon in the climate system: A scientific assessment. *J. Geophys. Res.* **2013**, *118*, 5380–5552.
4. Twohy, C.H.; Poellot, M.R. Chemical characteristics of ice residual nuclei in anvil cirrus clouds: Evidence for homogeneous and heterogeneous ice formation. *Atmos. Chem. Phys.* **2005**, *5*, 2289–2297.
5. Mordukhovich, I.; Wilker, E.; Suh, H.; Wright, R.; Sparrow, D.; Vokonas, P.S.; Schwartz, J. Black carbon exposure, oxidative stress genes, and blood pressure in a repeated-measures study. *Environ. Health Perspect.* **2009**, *117*, 1767–1772.
6. Cao, J.J.; Xu, H.M.; Xu, Q.; Chen, B.H.; Kan, H.D. Fine particulate matter constituents and cardiopulmonary mortality in a heavily polluted Chinese city. *Environ. Health Perspect.* **2012**, *120*, 373–378.

7. Shindell, D.; Kuylenstierna, J.C.I.; Vignati, E.; van Dingenen, R.; Amann, M.; Klimont, Z.; Anenberg, S.C.; Muller, N.; Janssens-Maenhout, G.; Raes, F.; *et al.* Simultaneously mitigating near-term climate change and improving human health and food security. *Science* **2012**, *335*, 183–189.

8. Cao, J.J.; Wang, Q.Y.; Chow, J.C.; Watson, J.G.; Tie, X.X.; Shen, Z.X.; Wang, P.; An, Z.S. Impacts of aerosol compositions on visibility impairment in Xi'an, China. *Atmos. Environ.* **2012**, *59*, 559–566.

9. Wang, Q.Y.; Cao, J.J.; Tao, J.; Li, N.; Su, X.L.; Chen, L.A.; Wang, P.; Shen, Z.X.; Liu, S.X.; Dai, W.T. Long-term trends in visibility and at Chengdu, China. *PLoS One* **2013**, *8*, e68894.

10. Huang, R.-J.; Zhang, Y.; Bozzetti, C.; Ho, K.-F.; Cao, J.-J.; Han, Y.; Daellenbach, K.R.; Slowik, J.G.; Platt, S.M.; Canonaco, F.; *et al.* High secondary aerosol contribution to particulate pollution during haze events in China. *Nature* **2014**, *514*, 218–222.

11. Lei, Y.; Zhang, Q.; He, K.; Streets, D. Primary anthropogenic aerosol emission trends for China. *Atmos. Chem. Phys.* **2011**, *11*, 931–954.

12. Han, S.; Kondo, Y.; Oshima, N.; Takegawa, N.; Miyazaki, Y.; Hu, M.; Lin, P.; Deng, Z.; Zhao, Y.; Sugimoto, N.; *et al.* Temporal variations of elemental carbon in Beijing. *J. Geophys. Res.* **2009**, *114*, D23202.

13. Cao, J.J.; Tie, X.X.; Xu, B.Q.; Zhao, Z.Z.; Zhu, C.S.; Li, G.H.; Liu, S.X. Measuring and modeling black carbon (BC) contamination in the SE Tibetan Plateau. *J. Atmos. Chem.* **2010**, *67*, 45–60.

14. Schwarz, J.P.; Spackman, J.R.; Fahey, D.W.; Gao, R.S.; Lohmann, U.; Stier, P.; Watts, L.A.; Thomson, D.S.; Lack, D.A.; Pfister, L.; *et al.* Coatings and their enhancement of black carbon light absorption in the tropical atmosphere. *J. Geophys. Res.* **2008**, *113*, D03203.

15. Shiraiwa, M.; Kondo, Y.; Moteki, N.; Takegawa, N.; Sahu, L.; Takami, A.; Hatakeyama, S.; Yonemura, S.; Blake, D. Radiative impact of mixing state of black carbon aerosol in Asian outflow. *J. Geophys. Res.* **2008**, *113*, D24210.

16. Gysel, M.; Laborde, M.; Olfert, J.S.; Subramanian, R.; Gröhn, A.J. Effective density of aquadag and fullerene soot black carbon reference materials used for SP2 calibration. *Atmos. Measur. Tech.* **2011**, *4*, 2851–2858.

17. Kondo, Y.; Sahu, L.; Moteki, N.; Khan, F.; Takegawa, N.; Liu, X.; Koike, M.; Miyakawa, T. Consistency and traceability of black carbon measurements made by laser-induced incandescence, thermal-optical transmittance, and filter-based photo-absorption techniques. *Aerosol Sci. Technol.* **2011**, *45*, 295–312.

18. Cao, J.J.; Wu, F.; Chow, J.C.; Lee, S.C.; Li, Y.; Chen, S.W.; An, Z.S.; Fung, K.K.; Watson, J.G.; Zhu, C.S.; *et al.* Characterization and source apportionment of atmospheric organic and elemental carbon during fall and winter of 2003 in Xi'an, China. *Atmos. Chem. Phys.* **2005**, *5*, 3127–3137.

19. Cao, J.J.; Zhu, C.S.; Chow, J.C.; Watson, J.G.; Han, Y.M.; Wang, G.; Shen, Z.; An, Z.S. Black carbon relationships with emissions and meteorology in Xi'an, China. *Atmos. Res.* **2009**, *94*, 194–202.

20. Gao, R.S.; Schwarz, J.P.; Kelly, K.K.; Fahey, D.W.; Watts, L.A.; Thompson, T.L.; Spackman, J.R.; Slowik, J.G.; Cross, E.S.; Han, J.H.; *et al.* A novel method for estimating light-scattering properties of soot aerosols using a modified single-particle soot photometer. *Aerosol Sci. Technol.* **2007**, *41*, 125–135.

21. Moteki, N.; Kondo, Y. Effects of mixing state on black carbon measurements by laser-induced incandescence. *Aerosol Sci. Technol.* **2007**, *41*, 398–417.

22. Wang, Q.Y.; Huang, R.-J.; Cao, J.J.; Han, Y.M.; Wang, G.H.; Li, G.H.; Wang, Y.C.; Dai, W.T.; Zhang, R.J.; Zhou, Y.Q. Mixing state of black carbon aerosol in a heavily polluted urban area of China: Implications for light absorption enhancement. *Aerosol Sci. Technol.* **2014**, *48*, 689–697.

23. Schwarz, J.P.; Gao, R.S.; Fahey, D.W.; Thomson, D.S.; Watts, L.A.; Wilson, J.C.; Reeves, J.M.; Darbeheshti, M.; Baumgardner, D.G.; Kok, G.L.; *et al.* Single-particle measurements of midlatitude black carbon and light-scattering aerosols from the boundary layer to the lower stratosphere. *J. Geophys. Res.* **2006**, *111*, D16207.

24. Stephens, M.; Turner, N.; Sandberg, J. Particle identification by laser-induced incandescence in a solid-state laser cavity. *Appl. Opt.* **2003**, *42*, 3726–3736.

25. Slowik, J.G.; Cross, E.S.; Han, J.H.; Davidovits, P.; Onasch, T.B.; Jayne, J.T.; Williams, L.R.; Canagaratna, M.R.; Worsnop, D.R.; Chakrabarty, R.K.; *et al.* An inter-comparison of instruments measuring black carbon content of soot particles. *Aerosol Sci. Technol.* **2007**, *41*, 295–314.

26. Moteki, N.; Kondo, Y. Dependence of laser-induced incandescence on physical properties of black carbon aerosols: Measurements and theoretical interpretation. *Aerosol Sci. Technol.* **2010**, *44*, 663–675.

27. Laborde, M.; Mertes, P.; Zieger, P.; Dommen, J.; Baltensperger, U.; Gysel, M. Sensitivity of the single particle soot photometer to different black carbon types. *Atmos. Measur. Tech.* **2012**, *5*, 1031–1043.

28. Huang, X.F.; Sun, T.L.; Zeng, L.W.; Yu, G.H.; Luan, S.J. Black carbon aerosol characterization in a coastal city in South China using a single particle soot photometer. *Atmos. Environ.* **2012**, *51*, 21–28.

29. Moteki, N.; Kondo, Y.; Miyazaki, Y.; Takegawa, N.; Komazaki, Y.; Kurata, G.; Shirai, T.; Blake, D.; Miyakawa, T.; Koike, M. Evolution of mixing state of black carbon particles: Aircraft measurements over the western Pacific in March 2004. *Geophys. Res. Lett.* **2007**, *34*, L11803.

30. Huang, X.F.; He, L.Y.; Xue, L.; Sun, T.L.; Zeng, L.W.; Gong, Z.H.; Hu, M.; Zhu, T. Highly time-resolved chemical characterization of atmospheric fine particles during 2010 Shanghai World Expo. *Atmos. Chem. Phys.* **2012**, *12*, 4897–4907.

31. Okuda, T.; Matsuura, S.; Yamaguchi, D.; Umemura, T.; Hanada, E.; Orihara, H.; Tanaka, S.; He, K.; Ma, Y.; Cheng, Y.; *et al.* The impact of the pollution control measures for the 2008 Beijing Olympic Games on the chemical composition of aerosols. *Atmos. Environ.* **2011**, *45*, 2789–2794.

32. Schwarz, J.P.; Stark, H.; Spackman, J.R.; Ryerson, T.B.; Peischl, J.; Swartz, W.H.; Gao, R.S.; Watts, L.A.; Fahey, D.W. Heating rates and surface dimming due to black carbon aerosol absorption associated with a major U.S. city. *Geophys. Res. Lett.* **2009**, *36*, L15807.

33. Wang, Q.Y.; Schwarz, J.P.; Cao, J.J.; Gao, R.S.; Fahey, D.W.; Hu, T.F.; Huang, R.-J.; Han, Y.M.; Shen, Z.X. Black carbon aerosol characterization in a remote area of Qinghai-Tibetan Plateau, western China. *Sci. Total Environ.* **2014**, *479*, 151–158.

34. Huang, X.F.; Gao, R.S.; Schwarz, J.P.; He, L.Y.; Fahey, D.W.; Watts, L.A.; McComiskey, A.; Cooper, O.R.; Sun, T.L.; Zeng, L.W.; *et al.* Black carbon measurements in the Pearl River Delta region of China. *J. Geophys. Res.* **2011**, *116*, D12208.

35. Schwarz, J.P.; Gao, R.S.; Spackman, J.R.; Watts, L.A.; Thomson, D.S.; Fahey, D.W.; Ryerson, T.B.; Peischl, J.; Holloway, J.S.; Trainer, M.; *et al.* Measurement of the mixing state, mass, and optical size of individual black carbon particles in urban and biomass burning emissions. *Geophys. Res. Lett.* **2008**, *35*, L13810.

36. Moffet, R.C.; Prather, K.A. *In-situ* measurements of the mixing state and optical properties of soot with implications for radiative forcing estimates. *Proc. Natl. Acad. Sci. USA* **2009**, *106*, 11872–11877.

37. Liggio, J.; Gordon, M.; Smallwood, G.; Li, S.M.; Stroud, C.; Staebler, R.; Lu, G.; Lee, P.; Taylor, B.; Brook, J.R. Are emissions of black carbon from gasoline vehicles underestimated? Insights from near and on-road measurements. *Environ. Sci. Technol.* **2012**, *46*, 4819–4828.

38. Kondo, Y.; Komazaki, Y.; Miyazaki, Y.; Moteki, N.; Takegawa, N.; Kodama, D.; Deguchi, S.; Nogami, M.; Fukuda, M.; Miyakawa, T.; *et al.* Temporal variations of elemental carbon in Tokyo. *J. Geophys. Res.* **2006**, *111*, D12205.

39. Baumgardner, D.; Raga, G.; Peralta, O.; Rosas, I.; Castro, T.; Kuhlbusch, T.; John, A.; Petzold, A. Diagnosing black carbon trends in large urban areas using carbon monoxide measurements. *J. Geophys. Res.* **2002**, doi:10.1029/2001JD000626.

40. Dickerson, R.R.; Andreae, M.O.; Campos, T.; Mayol-Bracero, O.L.; Neusuess, C.; Streets, D.G. Analysis of black carbon and carbon monoxide observed over the Indian Ocean: Implications for emissions and photochemistry. *J. Geophys. Res.* **2002**, doi:10.1029/2001JD000501.

41. Spackman, J.R.; Schwarz, J.P.; Gao, R.S.; Watts, L.A.; Thomson, D.S.; Fahey, D.W.; Holloway, J.S.; de Gouw, J.A.; Trainer, M.; Ryerson, T.B. Empirical correlations between black carbon aerosol and carbon monoxide in the lower and middle troposphere. *Geophys. Res. Lett.* **2008**, *35*, L19816.

42. Pna, X.L.; Kanaya, Y.; Wang, Z.F.; Liu, Y.; Pochanart, P.; Akimoto, H.; Sun, Y.L.; Dong, H.B.; Li, J.; Irie, H.; *et al.* Correlation of black carbon aerosol and carbon monoxide in the high-altitude environment of Mt. Huang in Eastern China. *Atmos. Chem. Phys.* **2011**, *11*, 9735–9747.

43. McMeeking, G.; Hamburger, T.; Liu, D.; Flynn, M.; Morgan, W.; Northway, M.; Highwood, E.; Krejci, R.; Allan, J.; Minikin, A.; *et al.* Black carbon measurements in the boundary layer over western and northern Europe. *Atmos. Chem. Phys.* **2010**, *10*, 9393–9414.

44. Baumgardner, D.; Kok, G.; Raga, G. On the diurnal variability of particle properties related to light absorbing carbon in Mexico City. *Atmos. Chem. Phys.* **2007**, *7*, 2517–2526.

45. Verma, R.L.; Sahu, L.K.; Kondo, Y.; Takegawa, N.; Han, S.; Jung, J.S.; Kim, Y.J.; Fan, S.; Sugimoto, N.; Shammaa, M.H.; *et al.* Temporal variations of black carbon in Guangzhou, China, in summer 2006. *Atmos. Chem. Phys.* **2010**, *10*, 6471–6485.

6

Wind Regimes above and below a Temperate Deciduous Forest Canopy in Complex Terrain: Interactions between Slope and Valley Winds

Xingchang Wang [1,*], Chuankuan Wang [1] and Qinglin Li [2,3]

[1] Center for Ecological Research, Northeast Forestry University, 26 Hexing Road, Harbin 150040, China; E-Mail: wangck-cf@nefu.edu.cn

[2] School of Forestry and Bio-technology, Zhejiang A & F University, 88 North Road of Huancheng, Lin'an 311300, China; E-Mail: geoliqinglin@yahoo.com

[3] Forest Analysis and Inventory Branch, Ministry of Forests, Lands, and Natural Resource Operations, 727 Fisgard Street, Victoria, BC V8W-9C2, Canada

* Author to whom correspondence should be addressed; E-Mail: xcwang_cer@nefu.edu.cn

Academic Editor: Albert A. M. Holtslag

Abstract: The thermally driven wind over mountainous terrains challenges the estimation of CO_2 exchange between forests and the atmosphere when using the eddy covariance technique. In this study, the wind regimes were investigated in a temperate deciduous forested valley at the Maoershan site, Northeast China. The wind direction above the canopy was preferentially up-valley in the daytime and down-valley in the nighttime, corresponding to the diurnal patterns of above-canopy temperature gradient and stability parameter. In both leaf-on and -off nighttime, a down-valley flow with a maximum velocity of 1~3 $m \cdot s^{-1}$ was often developed at 42 m above the ground (2.3-fold of the canopy height). However, the below-canopy prevailing wind was down-slope in the night, contrast to the below-canopy temperature lapse and unstable conditions. This substantial directional shear illustrated shallow slope winds were superimposed on larger-scale valley winds. As a consequence, the valley-wind component becomes stronger with increasing height, indicating a clear confluence of drainage flow to the valley center. In the daytime, the below-canopy wind was predominated down-slope due to the temperature inversion and stable conditions in the leaf-on season, and was mainly up-valley or down-slope in the leaf-off season. The isolation of momentum flux and radiation by the dense canopy played a key role in the formation of the below-canopy unaligned wind and inverse stability. Significant lateral kinematic momentum fluxes were detected due to the directional shear. These findings suggested a significant interaction between slope and valley winds at this site. The frequent vertical convergence / divergence above the canopy and horizontal divergence/convergence below the canopy in the nighttime / daytime is likely to induce significant advections of trace gases and energy flux.

Keywords: mountain wind system; valley terrain; directional shear; forest canopy; temperature gradient; eddy covariance

1. Introduction

The eddy covariance technique (EC) has been widely used to calculate the long-term net CO_2 exchange (NEE) between the atmosphere and terrestrial ecosystems in the last two decades [1]. However, the accuracy of EC measurements of NEE has been questioned particularly over tall vegetation in complex terrains [2–7] and even over various forests with relative flat terrains [7,8]. One of the key factors that results in uncertainties in measuring scalar exchanges between the atmosphere and forests is the complexity of airflow above and within the forest canopy [4,9–11]. Understanding mechanisms of airflow above and within a forest canopy provides a basis for interpreting nighttime CO_2 fluxes above the forest canopy [10,12–14].

In mountainous regions, disproportionate heating and cooling between airs in the valley slopes and the surrounding area result in mountain-valley circulations [15]. The nocturnal flows are typically down-slope and/or down-valley due to a surface temperature inversion, while the daytime airflows are normally up-slope and/or up-valley resulting from radiative heating and temperature lapse conditions [15]. On relatively simple sloping topography, the strength of the drainage flows (down-slope and down-valley winds) depends on slope gradient, distance from the crest, atmospheric stability and canopy structure [9,16–19]; on complex topography, the strength may be influenced by interactions between slope winds and valley winds [20–23]. In typical valley terrains, as the air cools down due to a negative radiation budget on the slope, a temperature inversion layer near the ground is developed before sunset. The cool air at the sidewalls moves downward along the slope, and then these flows converge on the valley floor with the flows downward along the valley axis [21]. Because the down-slope and down-valley winds are typically in orthogonal directions in most valleys, there is often a considerable directional shear through the slope wind layer [20]. In the daytime after sunrise, the radiative heating gradually breaks down the temperature inversion and then reverses the down-slope (down-valley) wind to up-slope (up-valley) wind [15]. Such wind regimes result in the complex nature of scalar transport, and make EC measurements in hilly topography difficult [11,24,25].

Forests, with tall canopies, are often distributed in mountainous terrains, which lead to a more complex wind regime than short vegetations (e.g., grassland, cropland) [12,18,26]. When a slope is covered by high and dense forest, the radiating surface to the night sky turns upward to the canopy, rather than the ground [27], and even an opposite temperature gradient is formed below canopy compared with that above canopy [24,28,29]. These situations are quite different with those of open canopy forest or bare ground. Beneath such dense and high forest canopies, persistent subcanopy nighttime upward and daytime downward flow patterns are observed [10,30], which contradicts the typical thermally driven winds [15]. Another distinct feature of dense canopy wind is the core of drainage flow may be beyond the canopy height [29,31]. However, the effects of the density of the canopy on generating a decoupling subcanopy layer and elevating the airflow are still not well known.

Most studies on wind regimes focus on nocturnal flows particularly in the leaf-on season, and few account for seasonal variations in thermally driven flows at small scales [10,31,32]. Daytime coupling/decoupling of airflows above and below forest canopy was investigated for a few studies [10,13]. However, because of the site-specific wind regime, it needs more researches to gain a general perspective on canopy winds in a complex topography. The combination of slope flows on the sidewalls and boundary layer processes on and above the valley floor is likely to cause valley winds [15], but the interaction between slope and valley winds in a forested area is poorly understood [10,22,30,33].

At the Maoershan Forest Ecosystem Research Station, a site of ChinaFLUX in Chinese temperate forests, Jiao et al [34] previously found that the daytime up-valley and nocturnal down-valley winds above the canopy had a classic characteristics of thermally driven winds. Accumulation of CO_2 under canopy at night was frequently observed during the summer. However, daytime temperature inversion under the dense forest canopy occurred frequently at this site [34]. Therefore, we are attempted to tackle the following questions: how do the wind regimes (*i.e.*, wind directions and velocities) in this forested valley vary vertically during the daytime and nighttime in both leaf-on and leaf-off seasons? Are there significant interactions between slope and valley winds at this site? Are the wind regimes influenced by the buoyancy forcing and atmospheric stability? What implications are the wind systems above and within the forest canopy to the EC measurements of NEE in complex terrains? To address these questions, we used the wind and temperature data on the 48-m-high flux tower at the Maoershan site to explore the diurnal variations and vertical shears of the canopy air flows in both nighttime and daytime during leaf-on and leaf-off seasons.

2. Slope/Valley Wind Theory

The slope wind system is a diurnal thermally driven wind system that blows up or down the slopes of a valley sidewall or an isolated hill or mountain, with upslope flows during daytime and down-slope flows during nighttime [15]. Slope flows result from the cooling or heating of an air layer over the slope relative to air at the same elevation away from the slope. The velocity of such thermotopographic flows depends largely on the scales of the flow and the forcing [10]. The drivers of thermotopographic flows may be examined using the equations of motion and the heat equation following Mahrt [35] and Zardi and Whiteman [15]. Here, we cite the three equations to provide a framework for our arguments. These equations are expressed base on a coordinate system that is oriented along (*s*) and perpendicular (*n*) to an infinitely extended slope of uniform inclination angle α [35]. *n* is perpendicular to the slope and increases upwards; *s* is parallel to the slope and increases in the down-slope direction. Using the Boussinesq approximation, the along-slope and slope perpendicular equations of motion, the thermodynamic energy equation, and the continuity equation are then written as

$$\frac{\partial u}{\partial t} + u\frac{\partial u}{\partial s} + w\frac{\partial u}{\partial n} = -\frac{1}{\rho_0}\frac{\partial (P - P_a)}{\partial s} - g\frac{\theta - \theta_0}{\theta_0}\sin\alpha - \frac{\partial \overline{u'w'}}{\partial n} \tag{1}$$

$$\frac{\partial w}{\partial t} + u\frac{\partial w}{\partial s} + w\frac{\partial w}{\partial n} = -\frac{1}{\rho_0}\frac{\partial (P - P_a)}{\partial n} + g\frac{\theta - \theta_0}{\theta_0}\cos\alpha \tag{2}$$

$$\frac{\partial \theta}{\partial t} + u\frac{\partial \theta}{\partial s} + w\frac{\partial \theta}{\partial n} = -\frac{1}{\rho_0 c_p}\frac{\partial R}{\partial n} - \frac{\partial \overline{w'\theta'}}{\partial n} \tag{3}$$

where u and w are the velocity components parallel to s and n, respectively. ρ is air density, and P is barometric pressure. θ is a reference value for the virtual potential temperature at the ground in the absence of surface cooling or heating, g is the gravitational acceleration, c_p is the specific heat of air at constant pressure, R is the upward radiative flux, and $\overline{u'w'}$ and $\overline{w'\theta'}$ are the kinematic turbulent Reynolds fluxes of momentum and heat. Variables with subscript "0" denote the mean conditions in the absence of slope flow, unsubscripted variables denote locally perturbed mean conditions associated with cooling/warming within the slope flow layer, and primes denote turbulent fluctuations around a local mean. The ambient pressure P_a is determined from hydrostatic balance. In this framework, it is assumed that the ambient air, in the absence of thermally driven flow, is at rest. Note that, consistent with the assumption of a shallow layer, equation (1) includes only the slope-normal turbulent fluxes, and lateral momentum fluxes are ignored. For simplicity, latent heat fluxes are not included here, although they can sometimes play a crucial role in the overall energy budgets. Accelerations normal to the slope are generally considered negligible so that the atmosphere normal to the slope is in quasi-hydrostatic balance and the RHS of Equation (2) is practically zero.

It is instructive to consider which terms in these equations are most important for understanding slope flow physics. Down-slope flows are not governed by slope angle (gravity) alone, but controlled synergistically by slope cooling, ambient stratification and vegetation structure [17]. Although the most important factor of optimal down-slope wind is slope angle [17], for a given site (the slope angle is fixed), the key term that drives slope flows is the buoyancy caused by the temperature excess or deficit that forms over the slope (second term on the right hand side of Equation (1)) [15].

Valleys, with geometries more complex than simple slopes, exhibit that the radiation patterns and thermally driven winds are consequently more complex. Unlike slope winds, diurnal valley winds are thermally driven that blow along the axis of a valley, with up-valley flows during daytime and down-valley flows during nighttime [15]. Valley winds are not primarily a function of the slope of the underlying valley floor. Instead, they depend on other geometrical factors, such as the shape and aspect of the valley cross-section and their along-valley variations [15]. The dynamics of diurnal valley winds results from the combination of slope flows on the sidewalls and boundary layer processes on and above the valley floor, including such factors as free convection, shear, and compensatory subsidence in the valley core atmosphere.

The evolution of real thermally driven winds is usually determined by the topographic setting and the flow and energy budget at a larger scale [22,36]. Highly complex terrain can be viewed as a series of valleys and slopes, with small slopes converging into small valleys or basins, which in turn merge to form larger and larger scale slopes or valleys [10,21]. In such topography, thermally driven winds may be generated at many different scales, with flow in the thin layer nearest to the ground responding to the local slope and local forcings and thicker flows responding to large-scale slopes and large-scale forcings [22,36]. Thermal and dynamical properties of flows at any scale influence flows on both larger and smaller scales [37]. Therefore, dynamic interactions between the slope and valley flows are typically observed on valley sidewalls [15].

Thermotopographic flows in vegetated terrain (particularly tall forested terrain) are likely more complex than flows in unvegetated topography. In forested complex terrain, we must consider not only radiative heating/cooling near the ground, but also at the top of the canopy [27]. There may be multiple layers of thermally driven flow, which are decoupled from each other [9,38,39]. In deciduous forests, the influences of canopy on drainage flow may be expected to be different in foliated and leaf-off seasons [10,39]. Here, we will explore the patterns of and the forcings of the wind systems at a valley sidewall site, in both leaf-on and leaf-off seasons.

3. Experimental Section

3.1. Site Description

This study was conducted at the Maoershan Forest Ecosystem Research Station of Northeast Forestry University, Heilongjiang Province, Northeast China (45°24′N, 127°40′E, 400 m a.s.l.). The climate is a continental monsoon climate with a windy and dry spring, a warm and humid summer, and a dry and cold winter. The mean (1989–2009) annual precipitation is 629 mm, of which ~50% falls between June and August. The mean annual air temperature is 3.1 °C, and mean January and July air temperature are −18.5 °C and +22.0 °C, respectively [40]. The frost-free period is between 120 and 140 days with early frosts in September and late frosts in May. The mean slope around the tower was ~9°, and the slope of the along-valley center was ~1°.

The vegetation is a naturally generated temperate deciduous broadleaved forest. The dominant tree species include *Fraxinus mandshurica* Rupr., *Ulmus japonica* Sarg., *Betula platyphylla* Suk., *Populus davidiana* Dode, *Juglans mandshurica* Maxim., *etc.* The stand was about 60-year-old, with a basal area of 26.7 $m^2 \cdot ha^{-1}$ and a canopy height of 18 m. The maximum leaf area index (LAI) was 5.7 $m^2 \cdot m^{-2}$ based on the litter fall collection method in eight permanent plots (Wang, unpublished data). Briefly, the LAI was calculated as the species- and time-specific leaf mass per area, and accumulated from August to October. A 1~8 m high and well-developed understory is dominated by *Syinga reticulata* var. *mandshurica*, which contributed 16% to 42% of the stand LAI (with a mean of 27%).

3.2. Measurements and Instrumentations

A 48-m-high eddy flux tower is set up at the toe of the sidewall (NW-facing) of a NE-SW towards-valley that is about 2 km wide and 240 m deep (Figure 1). The valley axis has an azimuth of 20° to north and 240° to southwest. An open path EC system (CSAT3, Campbell Scientific, USA and LI-7500, LI-COR, USA) was installed at 36 m height to measure the momentum, heat and CO_2 fluxes between forest and the atmosphere. Horizontal wind velocities, wind directions, and vertical wind velocities were measured at seven, three, and two height levels of the tower, respectively (Table 1). The winds at 36 m and 2 m were measured by ultrasonic anemometer at high frequency (10 HZ), wind speed at other levels (all over canopy) were measured by cup anemometers, with the lower wind speed threshold of 0.22 m·s^{-1}. To calculate the profile of the virtual potential temperature, air temperature and humidity were measured at five heights. The barometric pressure was measured at 28 m above the ground.

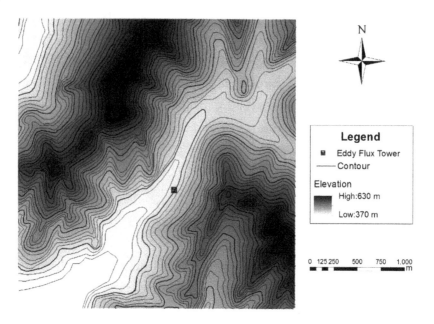

Figure 1. Contour map of the site.

Table 1. Instrumentation and measured variables at the flux tower.

Variable	Height (m)	Anemometer Type	Instrument, Supplier
Horizontal wind velocity	48, 42, 36, 28, 21, 16	Cup	010C, Met ONE
	2	Sonic	81,000, R.M. Yong
Horizontal wind direction	48	Vane	010C, Met ONE
	21	Vane	020C, Met ONE
	2	Sonic	R.M Young 81,000, R.M. Yong
Vertical wind velocity	36	Sonic	CSAT3, Campbell Scientific
Net radiation	48, 2	-	CRN1, Kipp & Zone,
Air temperature and humidity	48, 36, 28, 16, 2	-	HMP45C with 076B,Vessla
Barometric pressure	28	-	CS100, Campbell Scientific

3.3. Data Analysis

Four sectors of wind direction were divided: down-valley sector (from 350° to 50°), down-slope sector (from 50° to 210°), up-valley sector (from 210° to 270°) and up-slope sector (from 270° to 350°) according to the terrain. Then three conditions were considered as criteria of nocturnal above-canopy cold air drainage: (1) the wind direction must be within ±30° downward along the valley-axis (from 350° to 50°) [38]; (2) the horizontal wind-velocity profile must have a local maximum near the surface (usually at 42 m in this study) [15]; and (3) the nighttime was defined as that the global radiation measured at 48 m was less than 20 W·m^{-2} [39].

Valley wind components were projected to the along-valley and cross-valley directions: the velocity components along the valley axis, consisted of down-valley component (negative value) and up-valley component (positive value); down-slope component (negative value) and up-slope component (positive value). The thermally driven flow intensity was expressed with the velocity components. The vertical shear of horizontal velocity just above the height of the drainage center ($U_{48m} - U_{42m}$)/(48 m − 42 m) was also a useful measure of the thermally driven wind flow, where U was the horizontal wind velocity. Moreover, the vertical velocity, corrected with the planar-fit method [41], was another measure of thermally driven wind [4,10].

The basic physical processes in the lowest atmospheric layers are the downward transport processes of momentum and the dissipative processes of turbulent energy cascades [42]. Drag, generated when a fluid moves over the ground or through vegetation, is essential in this process. Drag creates velocity gradients and eddies, characterized by the profiles of flow velocity and Reynolds stress, which lead to momentum loss of the fluid [42]. Drag coefficient (C_D) related to friction velocity and wind velocity square [43,44],

$$c_D(z) = \frac{u_*^2(z)}{\overline{U}^2(z)} \tag{4}$$

where $u_*(z)$ is the friction velocity at height z and was related to the shear stress, U is the speed of the wind vector measured by sonic anemometer after tilt-correction by planar fit method. The physical meaning of $c_D(z)$ is the effectiveness of canopy drag elements in absorbing momentum from the airflow [42]. Unlike at simple slope of an isolated mountain, the air flow in the valley sidewall may have distinct directional shear [20,33], so we adopted the longitudinal and lateral shear stress in the u_* [33,44,45] after planar fit correction [41],

$$u_* = \left(\overline{u'w'}^2 + \overline{v'w'}^2 \right)^{1/4} \tag{5}$$

where u, v and w were the along-wind, crosswind and vertical wind components, respectively. The prime indicated instantaneous deviation from the mean, over bar meant time average (half-hour).

The relations between the wind regime and the forcing was firstly through the buoyancy effect [15], which was expressed with difference in virtual potential temperature (θ_v) [24] between two height levels. A positive temperature gradient ($\Delta\theta_v/\Delta z$) was indicative of a downward buoyancy force, and vice versa [24]. Because of the canopy isolation, we separately calculated the above canopy θ_v gradient (($\theta_{v48m} - \theta_{v16m}$)/(48 m − 16 m)) and the subcanopy θ_v gradient (($\theta_{v16m} - \theta_{v2m}$)/(16 m − 2 m)). We analyzed the relations of bin-averaged wind velocity to temperature gradients [9,41]. To gain further insight into the wind regime, the distributions of wind direction were plotted for various stability classes. The stability parameter $\zeta = -(z - d)/L$ was used in this study, where z was measurement height of EC (36 m), d was zero plane displacement (12.0 m for 36 m and 0.3 m for 2 m) and L was Monin-Obukhov length. The five stability regimes [28,46] have been defined as following: strongly unstable ($\zeta < -5$), moderately unstable ($-5 < \zeta < -0.05$), near-neutral ($-0.05 < \zeta < 0.05$), moderately stable ($0.05 < \zeta < 5$) and strongly stable ($\zeta > 5$). The ζ was calculated for both above-canopy and below-canopy layers. To further explore the effect of canopy density on canopy wind flows, we divided the calendar year into two seasons: the leaf-on season (11 May to 5 October) and leaf-off season (the other times). The subcanopy decoupling with that above the canopy was defined as contrasting momentum fluxes, opposite stability conditions and thermal gradients between layers. Generally, all of the three aspects of subanopy decoupling were related to a misalignment on wind direction between layers [9,10,18,38,39].

It was worth noting that this study focused on the general patterns of and the common forcing on the air flow above and below the forest canopy, so the temporal intermittency [23,28] and horizontal variability [7,42–44] were not emphasized.

4. Results and Discussions

4.1. Variations in Wind Direction

4.1.1. Diurnal Variations in Wind Direction

For the whole year, the wind directions above canopy (48 m and 21 m) were along-valley in a relative frequency of 80% (Table 2). During the nighttime, the winds were predominantly down-valley, and during the daytime the winds were preferentially up-valley (Table 2). Down-valley winds occurred at 60% to 80% of the times at night and diminished to about 20% in the day, whereas the up-valley winds showed an opposite diurnal pattern. The regular changes in wind direction indicated that the wind above the canopy was a thermally driven wind system.

Table 2. Frequency of down- and up-valley, and down- and up-slope winds under the canopy, just above canopy and above canopy in the daytime and nighttime during the leaf-on and leaf-off seasons of 2012. The data are calculated using 30-min average wind vectors. The dominant wind directions are bolded.

Measuring Height	Wind Direction	All Season	Leaf-on Season		Leaf-off Season	
			Day	Night	Day	Night
(m)	sector	(%)	(%)	(%)	(%)	(%)
48 (above-canopy)	Down-valley	**44.4**	30.1	**77.3**	21.5	**51.1**
	Down-slope	8.9	16.0	6.1	6.2	7.6
	Up-valley	41.2	**47.7**	14.7	**62.4**	37.1
	Up-slope	5.5	6.3	1.9	9.9	4.1
21 (just above canopy)	Down-valley	**45.6**	27.8	**73.6**	21.5	**58.1**
	Down-slope	10.3	11.5	14.4	6.2	10.0
	Up-valley	38.9	**50.2**	11.0	**63.8**	30.0
	Up-slope	5.3	10.5	1.0	8.6	2.0
2 (under-canopy)	Down-valley	21.7	7.7	29.5	9.5	35.6
	Down-slope	**53.4**	**56.7**	**62.7**	35.7	**57.1**
	Up-valley	14.3	15.2	2.3	**37.3**	5.4
	Up-slope	10.7	20.4	5.6	17.5	1.9

The wind direction below the canopy (2 m) was quite different from that above canopy (Table 2). Subcanopy prevailing direction was down-slope at all times, with the exception of a slightly higher frequency of up-valley wind in the leaf-off afternoon. The frequency of the subcanopy down-valley winds was <30% and decreased to its minimum (<10%) in the daytime. In contrast, the frequency of the subcanopy up-valley and up-slope winds increased slightly in the daytime. These patterns of subcanopy winds suggested a frequent misalignment between above- and below-canopy winds (See Section 4.1.2. Vertical Shear of Wind Direction for details).

Although the diurnal patterns were overall similar in the leaf-on and leaf-off seasons, there were still some differences. At the 48 m level, nocturnal down-valley winds trended to have a higher frequency during the leaf-on season, whereas the up-valley winds were more frequent in both daytime and nighttime during the leaf-off season. At the 21 m level, the down slope-winds were more frequent during the leaf-on season than during the leaf-off season. At the 2 m level, the up-valley winds were more frequent during the leaf-on season than during the leaf-off season (Figure 2c,f). These findings were overall consistent with previous studies that accounted of the seasonal variability of thermally driven winds [9,28,31,38].

4.1.2. Vertical Shear of Wind Direction

Distinct vertical shears of wind direction were observed during both daytime and nighttime (Table 2). In the daytime, the wind direction at 21 m was in consistent with that at 48 m, yet that at 2 m shifted counter-clockwise to down-slope direction. In addition, the daytime counter-clockwise and nighttime clockwise shears of wind directions were more dramatic during the leaf-on season than the leaf-off season (Table 2). The frequencies of down-slope winds increased 40.7% and 29.5% from 48 m to 2 m in leaf-on and leaf-off seasons, respectively. The frequencies of up-valley winds correspondingly decreased from 48 m to 2 m by 32.5% and 25.1% in leaf-on and leaf-off seasons. At night, the dominating winds sheared clockwise to more than 40° from 48 m to 2 m. Compared with those at 48 m, the percentages of down-valley winds at 2 m decreased 47.8% during the leaf-on season, while decreased only 15.5% during the leaf-off season. The relative frequencies of down-slope winds increased 56.6% and 49.5% for the leaf-on and leaf-off seasons, correspondingly (Table 2).

The right (clockwise) shear of subcanopy flow at night was in qualitative agreement with previously observed and modeled horizontal mean wind directional shear in some forest sites [33,45–48], but conflicted to the left shift in other sites [32,43,49]. In fact, the subcanopy wind direction in calm night was highly dependent on the local longest slope direction [18,36], which is highly local site-dependent. However, the upper valley wind system was determined by the larger scale topography [14,32]. At our site, the clear down-slope direction shear from above-canopy layer towards the ground surface during the night indicated a confluence of down-slope flows to the valley center [19,20].

Using the 20° threshold of wind direction shear between height levels after Alekseychik et al. [8], we found the above canopy misalignment had a probability of 22% in the daytime and 39% in the nighttime. However, the below-canopy misalignment occurred much more frequent, 76% in the daytime and 71% in the nocturnal occasions. On one hand, the nocturnal misalignment of the subcanopy flow with that above canopy is a common feature of a site with a tall and dense canopy in complex terrain [8,9,12,18,43,47,50], and even at a old-growth tropical forest site with flat terrain (Santarém LBA-ECO site, ~1° slope) [8]. On the other hand, the frequent subcanopy down-slope winds during the day (Table 2) contradicts with the classical theory of diurnal mountain wind systems [15]. The daytime drainage flows under tall and dense canopies were also observed at other sites: down-slope at sites with flat terrain [8,30] or down-valley at complex sites [9,28,51]. Interestingly, the five sites with a daytime drainage flow below canopy (the MMSF site [10], the AEF site [29], the Manaus LBA site [30], the Xishuangbanna site [47] and the Maoershan site) had two common features: (1) a tall, dense canopy; (2) closed to the valley center or at the lower part of the sidewall with relative gentle slope. There is evidence that the radiative heating/cooling of dense canopy isolated the subcanopy layer with that above canopy [26,28], which plays a key role in the formation of these decoupled subcanopy flows. However, it is unclear why the five cases of daytime subcanopy drainage all occurred near the center of a valley. At our site, the daytime subcanopy down-slope wind may be a partial compensation for the up-valley wind above canopy. The frequent subcanopy drainage flow in both daytime and nighttime might question the one-point eddy flux observation above canopy.

4.2. Variations in Wind Velocity

4.2.1. Horizontal Wind Velocity

The mean diurnal courses of horizontal wind velocities over the canopy had roughly bimodal patterns in the leaf-on season but near unimodal in the leaf-off season (Figure 2). The above-canopy horizontal wind velocity was lowest in the morning and evening transition periods. The morning transition period occurred at two to three hours after sunrise and the evening transition started about one hour before sunset. In contrast, the horizontal wind velocity at 2 m was overall higher in the nighttime than in the daytime, although it was generally lower than that above the canopy.

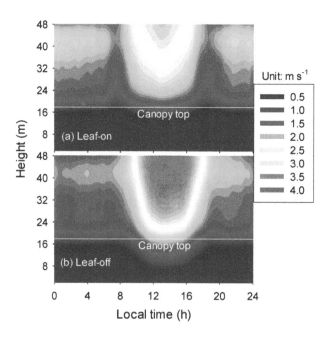

Figure 2. Diurnal courses of horizontal wind velocity profile during the leaf-on (**a**) and leaf-off (**b**) seasons of 2012.

The vertical profiles of mean horizontal velocities varied with the time of the day, although that above canopy was always higher than that below canopy (Figure 2). The horizontal wind velocity had a distinct maximum at the height of 42 m (2.3-fold canopy height) during night, while it increased with height during the daytime. However, in contrast to the higher mean wind velocity in leaf-off season, the wind maximum was stronger in leaf-on season, indicating the drainage flow was stronger in the leaf-on season.

Other researches documented that the typical height of nocturnal drainage center is 10 to 100 s of meters [15]. For example, the height of wind velocity maximum on the low-angle (1.6°) side-wall of the floor of Utah's Salt Lake Valley occurred at 10 to 15 m height [20], but could reach 40 m in another slope in the same valley [48]. Most studies in forest sites reported that the drainage flows were restricted in the subcanopy layer, which characterized with relative open canopy and insignificant understory vegetation [8,13,21], or at the upper part of a long slope [39,40], or very gentle slope [39]. We did not detect the possible subcanopy wind maximum as many previous studies [13,28,31,41,42] due to only one observation point below canopy.

The mean drainage maximum above canopy (42 m) at the Maoershan was relative high compared to other sites, which was likely the result of the valley terrain. In valley topography, significant interplays between the flows on side-wall and the flows along the valley axis, down-valley flows differ dramatically from down-slope flows on isolated mountains [15]. When the observation was conducted at upper slope, the depth of nocturnal drainage flow was often limited to the first few meters above the ground in the trunk-space e.g., [8,13,39,40]. On the side-wall of a valley, the down-slope flow often started before sunset [23,49], and then speeded up and increased in depth with down-slope distance before mid-night [18,19,23]. However, down-slope flows on valley sidewalls were routinely subject to the development of overlying valley flows [14,19]. In fact, down-slope flows typically converged onto the valley floor [20,45] or into the elevated portion of temperature inversion layer on the valley floor [20]. If the valley did not have major constrictions along its length or major changes in surface microclimates, the nighttime down-valley flows accelerated and increased in depth with down-valley distance [15]. So the main characteristic of the nighttime phase is the predominance of down-valley flows throughout the valley volume [15]. Unfortunately, most eddy flux towers were not tall enough to observe the jet-maximum of drainage flow above the canopy. Only a few studies (including this study) detected that the height of maximum wind velocity could occur over the canopy [21,28,30]. However, determining the whole profile of the valley-wind in any site needs high tower or other techniques (e.g., tethersonde balloon and sodar) [29]. The deep drainage in the night might question the mass and energy fluxes measured by EC above canopy [25].

4.2.2. Vertical Wind Velocity

The vertical wind velocity at 36 m (after planar-fit coordinate rotation) had a distinct diurnal pattern during the leaf-on season, with upward directions (positive values) in the daytime and downward directions (negative values) in the nighttime (Figure 3a). The vertical velocity at the 2 m was slightly downward at most times except for mid-morning and early-evening (Figure 3c). The nocturnal vertical velocity was in the order of -0.05 m·s^{-1}. The mean vertical velocities during the leaf-off season were near zero all day long (Figure 3b,d), although the variability between days was comparable with those during the leaf-on season.

The above-canopy diurnal pattern of daytime-upward and nighttime-downward vertical velocity, consistent with the diurnal course of wind direction, was a distinct feature of thermally driven wind system [3,4,18,24,45,52]. The leaf-on daily change in the above-canopy vertical velocity indicated that a convergence dominated in the nighttime and a divergence prevailed in the daytime [9,24,52]. This indicated positive vertical advection (removing CO_2 from the control volume), ignoring which in the mass balance equation would underestimate the nocturnal respiration [25]. The nocturnal above-canopy convergence was likely compensated by the below-canopy horizontal divergence, in view of the systematic directional shear to down-slope from 48 to 2 m. In this situation, the sign and magnitude of horizontal advection could not be determined without knowing the horizontal CO_2 gradient [25]. The positive vertical velocity in the daytime indicated a negative vertical advection, ignoring which would underestimate the CO_2 uptake by the ecosystem. However, the contribution of total vertical and horizontal advection to the NEE depends on their signs and magnitudes, which are difficult to infer without direct horizontal advection measurements.

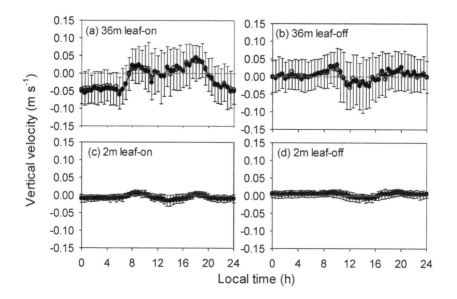

Figure 3. Diurnal courses of vertical wind velocity above (36 m) and below (2 m) the canopy during the leaf-on and leaf-off seasons of 2012. (**a**) 36 m for leaf-on season, (**b**) 36 m for leaf-off season, (**c**) 2 m for leaf-on season, and (**d**) 2 m for leaf-off season. Open circles are mean values, filled circles are the median and error bars indicate the range between the 25 and 75 percentile, respectively. The dashed lines are zero velocities.

4.3. Variations in Thermal Gradient and Stability

During the leaf-on season, the mean subcanopy θ_v gradient was nearly opposite to that above canopy: negative in the daytime and positive in the nighttime above the canopy (Figure 4a), but positive or neutral in the daytime and negative in the nighttime below the canopy (Figure 4c). Interestingly, the below-canopy θ_v gradient was bimodal, with the small negative "saddle" at early-afternoon (~13:00). On the contrary, during the leaf-off season the below-canopy temperature gradient was generally consistent with that above canopy: negative from mid-morning to late-afternoon (Figure 4b,d). The above-canopy temperature gradient generally agreed with the classic diurnal course. However, the daytime temperature inversion and nighttime temperature lapse or isothermal conditions had been observed under tall and dense forests [10,29,30,39,50,51,53], which differed markedly from open canopies or the leaf-off season [9,33]. In the daytime with clear sky, the heating mainly occurred at the top of the canopy because the dense canopy absorbed much more solar radiation than the ground

[28,46]. As illustrated in Figure 5, the net radiation was roughly mirrored the thermal gradient for the corresponding layer (Figure 4). Surprisingly, the below-canopy daytime net radiation peaked at the early-afternoon (~13:00) rather than at the noon, corresponding to the pattern of below-canopy thermal gradient (Figure 4c). This is caused by that on the northwest-facing slope, the solar elevation angle is still high at that time (high radiation above canopy), and meanwhile the shadow effect of canopy is relative small (the length of the pathway for radiation penetration in the canopy is relative small, which is largest when the sun is "behind" the slope and smallest when the sun is "facing" to the slope). However, quantification of this effect is beyond the scope of this paper.

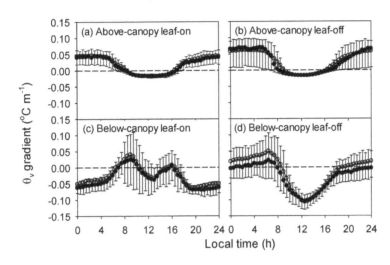

Figure 4. Diurnal courses of the gradients of virtual potential air temperature (θ_v) above (16 m–48 m) and bellow (2 m–16 m) the canopy during the leaf-on and leaf-off seasons of 2012. (**a**) above-canopy for leaf-on season, (**b**) above-canopy for leaf-off season, (**c**) below-canopy for leaf-on season, and (**d**) below-canopy for leaf-off season. Open circles are mean values, filled circles are the median and error bars indicate the range between the 25 and 75 percentile, respectively. The dashed lines are zero gradients.

The mean below-canopy daytime net radiation was comparable for the leaf-on and leaf-off seasons, although the above-canopy net radiation was higher during the leaf-on season. The canopy layer on average net absorbed 199 W·m^{-2} during the leaf-on daytime, whereas it decreased to 125 W·m^{-2} during the leaf-off daytime. This resulted in an above-canopy unstable but subcanopy thermally stable or near neutral stratification under dense canopy as shown in Figure 4 [9,12,29,47]. On the other hand, in clear and calm nighttime, canopy elements (the surfaces of leaves, branches and boles) initially cool much more quickly than the canopy air space after the net radiative cooling starts [27], leading to a cool center often at the top of canopy, which in turn resulted in a stable above-canopy layer but unstable subcanopy layer [9,12,29,47]. At our site, it was clear that the net radiation absorbed by the canopy was much stronger in the leaf-on season than in the leaf-off season, indicating a heat isolation of dense canopy (Figure 4). Notably, in a short and very dense conifer canopy (9.5 m high with a LAI of 9 m^2·m^{-2}), the temperature gradient within-canopy was positive [54]. It seemed that this short canopy had no sufficient space to form an unstable subcanopy layer because the foliage reached the ground. Therefore, a dense canopy and sufficient trunk space are indispensable requirements for a subcanopy inversion layer.

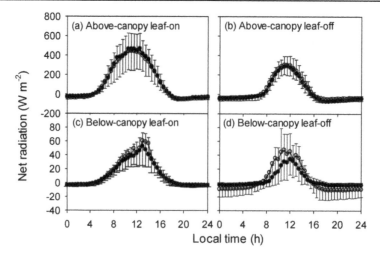

Figure 5. Diurnal courses of the net radiation above (48 m) and bellow (2 m) the canopy during the leaf-on and leaf-off seasons from May of 2013 to April of 2014. (**a**) 48 m for leaf-on season, (**b**) 48 m for leaf-off season, (**c**) 2 m for leaf-on season, and (**d**) 2 m for leaf-off season. Open circles are mean values, filled circles are the median and error bars indicate the range between the 25 and 75 percentile, respectively. The dashed lines are zero radiations.

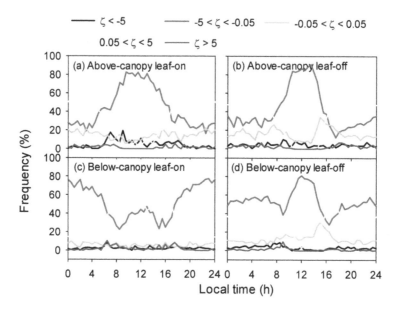

Figure 6. Diurnal courses of relative frequency of stability regimes for above-canopy (36 m) and below-canopy (2 m) layer in the daytime and nighttime during the leaf-on and leaf-off seasons of 2012. (**a**) 36 m for leaf-on season, (**b**) 36 m for leaf-off season, (**c**) 2 m for leaf-on season, and (**d**) 2 m for leaf-off season.

The frequencies of stability regimes were shown in Figure 6. At the 36 m height, the moderate unstable conditions dominated in the daytime and moderate stable conditions were prevailed during the night for both seasons. However, at the 2 m height, these situations were reversed, except for the daytime during the leaf-off season. This suggested the local stability of below-canopy layer was often opposite to that of above-canopy layer, which was due to the canopy isolation of vertical momentum and heat transport. Decoupling of above-canopy unstable and sub-canopy stable layers was also observed in relative open forest canopy at night, which was due to quite different mechanisms with that in dense canopy [6].

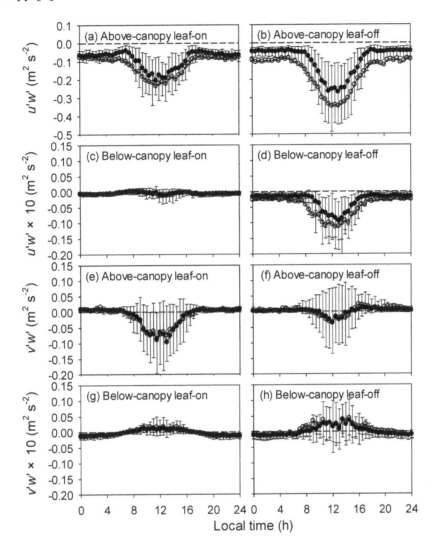

Figure 7. Diurnal courses of Reynolds stress components above (48 m) and bellow (2 m) the canopy during the leaf-on and leaf-off seasons of 2012. (**a**)–(**d**): longitudinal Reynolds stress, (a) 36 m for leaf-on season, (b) 36 m for leaf-off season, (c) 2 m for leaf-on season, and (d) 2 m for leaf-off season, (**e**)–(**h**): lateral Reynolds stress, (e) 36 m for leaf-on season, (f) 36 m for leaf-off season, (g) 2 m for leaf-on season, and (h) 2 m for leaf-off season. Open circles are mean values, filled circles are the median and error bars indicate the range between the 25 and 75 percentile, respectively. The dashed lines are zero gradient. Note the different vertical axis in the upper-most row of the panels.

4.4. Variations in Reynolds Stress and Drag Coefficient

Turbulence data are usually analyzed in a coordination system with along-wind, crosswind and vertical components in order to distinguish between the different characteristics of longitudinal and lateral turbulence. For the vertical momentum transport components of the Reynolds stress tensor, this means that in homogeneous terrain the lateral component is actually zero and the non-alignment between the mean horizontal wind vector and the stress vector is purely statistical [55]. In situations where two wind systems of different (spatial and time) scale are interacting this can of course not be expected [22]. The mean diurnal variations of above- and below-canopy Reynolds stress components (kinematic momentum fluxes) were shown in Figure 7. The longitudinal stress was negative and reached its negative maximum around the noon, roughly opposite to the daily course of horizontal wind velocity (Figure 2). The below-canopy longitudinal stress was much small in magnitude, and could be positive during the leaf-on season. Three distinct features could be detected for the lateral stress (Figure 4): (1) the magnitude of lateral stress was in the same order at, or even larger than (below canopy during the leaf-on season) the longitudinal one; (2) the daily course of longitudinal stress was similar between above and below canopy, whereas the pattern of lateral stress above canopy was reversed from that below canopy; (3) the lateral stress fluxes were generally small and of varying sign especially in the daytime. These findings illustrated a significant interaction between slope and valley winds.

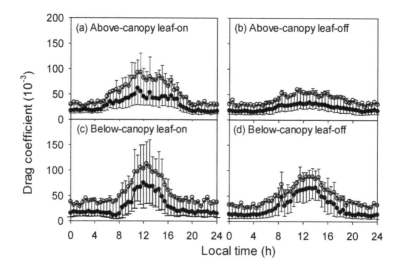

Figure 8. Diurnal courses of drag coefficients above (36 m) and bellow (2 m) the canopy during the leaf-on and leaf-off seasons of 2012. (**a**) above-canopy for leaf-on season, (**b**) above-canopy for leaf-off season, (**c**) below-canopy for leaf-on season, and (**d**) below-canopy for leaf-off season. Open circles are mean values, filled circles are the median and error bars indicate the range between the 25 and 75 percentile, respectively.

Previous momentum transport in canopy often simplified to a 2-D condition, *i.e.*, ignored the directional shear component of kinematic stress. Our results as well as the findings at other valley sites give the evidence that the lateral component of vertical momentum flux cannot be ignored. The opposite sign of below-canopy lateral stress with the above-canopy one at Maoershan site was in consistent with the findings above a steep forested slope in an alpine valley [22]. However, Rotach *et al.* [33] found that the lateral component of Reynolds stress exhibited a similar daily cycle in timing and magnitude as the component due to frictional (shear) stress at a valley sidewall, which was covered by meadow vegetation. It seems that the canopy layer played an essential role between the discrepancies. At our site, the upward moving fluctuations experienced a left, and the downward moving eddies a right turn, respectively. This situation seemed to be reversed at the 2 m height, indicating a decoupling subcanopy layer. However, detailed profile of stress measurement is necessary to give a whole picture of the momentum transport within the canopy and thus may help to interpret the site-specific regimes [22].

For the drag coefficient c_D, the below-canopy ones were slightly larger than the above-canopy ones particularly during the leaf-off season (Figure 8). This is due to the maximum c_D values are located around the maximum leaf area density levels for most forest canopies [42]. However, the c_D had a distinct daily course, which kept at a low level at night although the wind was continuous (Figure 2) and peaked around the noon when the wind speed reached its maximum (Figure 2). In addition, the daily course of c_D was similar to that of unstable condition (opposite to the stable condition), indicating a negative correlation between c_D and stability as previous findings [44,56]. Moreover, the c_D was larger in the leaf-on season than in the leaf-off season. These patterns of c_D indicated a nocturnal decoupling at the tower site, which could suppress momentum transport and scalar mixing during this period.

4.5. Relations to Thermal Gradient and Stability

4.5.1. Above-Canopy

The wind directions were related to the above-canopy θ_v gradient (Figure 9). Overall, in the conditions of moderate to strong above-canopy inversion (positive θ_v gradient), the down-valley wind occurred more frequently during both daytime and nighttime, whereas the down-slope wind became more important only during the day. These patterns illustrated that negative buoyancy forced the occurrence of drainage flow. Similarly, the stability strongly influenced the wind direction distribution (Figure 10). At 48 m and 21 m, the situation that the wind was dominated by up-valley direction at unstable conditions gradually converted to down-valley wind when conditions became moderately stable ($0.05 < \zeta < 0.5$). However, for strongly stable conditions, the wind changed to up-valley wind dominating again (Figure 10i). Interestingly, the down-slope wind prevailed at 2 m height level during the leaf-on season for all stability regimes, particular for strongly-unstable conditions (Figure 10a), whereas during the leaf-off season the down-slope wind became more frequent only when the conditions were stable (Figure 10h,j). The wind direction shift from down-valley at 48 m to down-slope at the 2 m was more dramatic at moderate stable conditions for both leaf-on and leaf-off seasons. The magnitude of directional shear was smallest for the near-neutral conditions, which was consistent with the situations at University of Michigan Biological Station site [55].

The velocities of the valley and slope winds were overall negatively correlated with the θ_v gradient (Figure 11). This suggested that the down-valley and down-slope winds were favored by the strong negative buoyancy (positive temperature inversion). However, there were distinct differences between valley and slope winds, between nighttime and daytime and between seasons. Compared with slope wind, the valley wind was more sensitive to the change of the above-canopy thermal gradient. At the same level of thermal inversion, the down-valley and down-slope winds were stronger in the nighttime than in the daytime. In addition, the down-valley wind was slightly stronger in the leaf-on season than in the leaf-off season.

The velocity shear between 42 and 48 m also responded strongly to the temperature gradient (Figure 12). During the nighttime, the negative velocity shear occurred when temperature inversion formed and reached the most negative values at moderate inversions, and then the velocity shear recovered to near zero when strong inversion formed (Figure 12a). This nonlinear response suggested that the strongest drainage flow occurred at moderate temperature inversion conditions, consistent with the response of the down-valley frequency (Figure 9a,d) and the along-valley velocity (Figure 11a). Country to the nighttime, the daytime velocity shear decreased gradually with the temperature gradient increasing (Figure 12b), roughly coincident with the trends of the down-valley frequency (Figure 9g) and the valley wind velocity (Figure 11b).

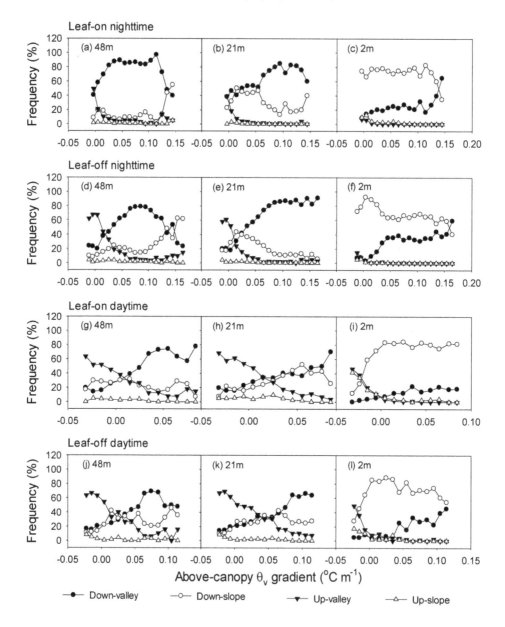

Figure 9. Changes of the relative frequency of mean wind direction sector at the 48 m, 21 m and 2 m with the gradient of above-canopy virtual potential air temperature (θ_v) in 2012. (**a**), (**b**), (**c**) are in the nighttime for 48 m, 21 m and 2 m in the leaf-on season, respectively, and (**d**), (**e**), (**f**) are in the nighttime for 48 m, 21 m and 2 m in the leaf-off season, correspondingly, (**g**), (**h**), (**i**) are in the daytime for 48 m, 21 m and 2 m in the leaf-on season, respectively, and (**j**), (**k**), (**l**) are in the daytime for 48 m, 21 m and 2 m in the leaf-off season, correspondingly.

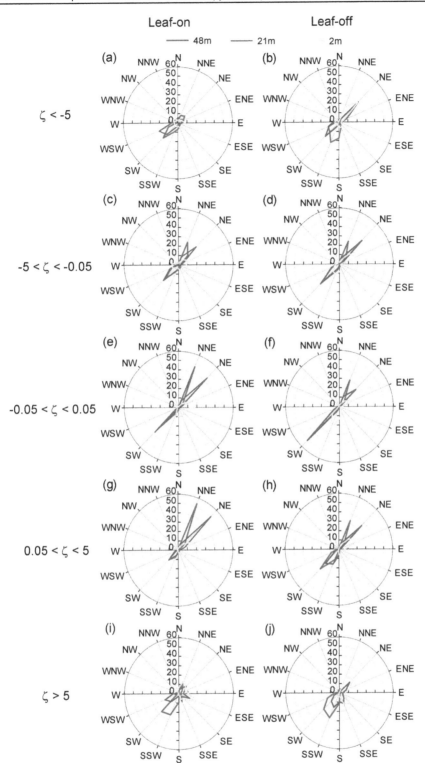

Figure 10. Relative frequency distributions of wind direction at the 48 m, 21 m and 2 m for five stability regimes during the leaf-on and leaf-off seasons of 2012. (**a**) and (**b**) are for $\zeta < -5$ in leaf-on and leaf-off seasons, correspondingly. (**c**) and (**d**) are for $-5 < \zeta < -0.05$ in leaf-on and leaf-off seasons, respectively. (**e**) and (**f**) are for $-0.05 < \zeta < 0.05$ in leaf-on and leaf-off seasons, respectively. (**g**) and (**h**) are for $0.05 < \zeta < 5$ in leaf-on and leaf-off seasons, respectively. (**i**) and (**j**) are for $\zeta > 5$ in leaf-on and leaf-off seasons, correspondingly.

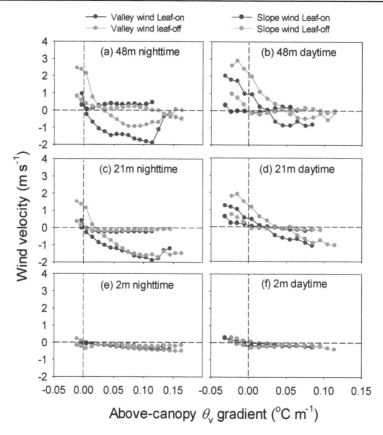

Figure 11. Changes of mean along-valley and cross-valley wind velocities at the 48 m, 21 m and 2 m with the gradient of above-canopy virtual potential air temperature (θ_v) in 2012. (**a**) and (**b**) are at 48 m in the nighttime and daytime, respectively. (**c**) and (**d**) are at 21 m in the nighttime and daytime, correspondingly. (**e**) and (**f**) are at 2 m in the nighttime and daytime, respectively.

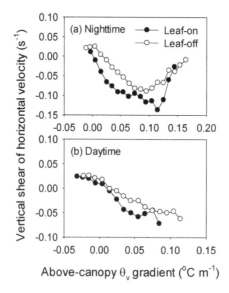

Figure 12. Changes of the vertical shear of horizontal wind speed between 42 and 48 m with the gradient of above-canopy virtual potential air temperature (θ_v) in the nighttime (**a**) and daytime (**b**) in 2012.

4.5.2. Below-Canopy

As expected, wind direction responded to the change in the below-canopy θ_v gradient (Figure 13). During the nighttime, the frequency of down-valley flow increased with the below-canopy θ_v gradient before a threshold and then trended to level off (Figure 13a,c). Surprisingly, the percentage of down-slope wind decreased under inversion conditions. During the daytime, the wind mainly blew to up-valley direction when the θ_v gradient above -0.1 °C·m^{-1}, the wind turned to down-slope direction when θ_v gradients were less inversed or strongly lapsed (Figure 13b,d). These responses with thermal gradient were roughly consistent with those with the stability regime (Figure 14). The down-slope wind dominated for all stability classes even for strongly unstable conditions during the leaf-on season (Figure 14a), with the exception of the strongly unstable and near-neutral conditions in the leaf-off season. The dominance of down-slope wind was most dramatic for strongly stable conditions under the canopy.

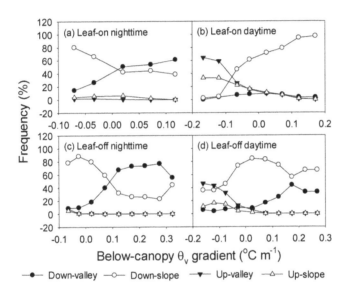

Figure 13. Change of the relative frequency of mean wind direction sector at the 2 m with the gradient of below-canopy virtual potential air temperature (θ_v) in 2012. (**a**) nighttime for leaf-on season, (**b**) daytime leaf-on season, (**c**) nighttime for leaf-off season, and (**d**) daytime for leaf-off season.

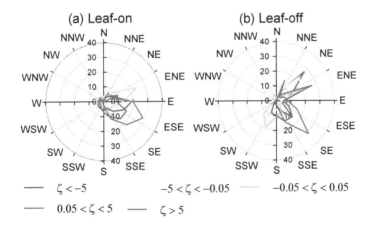

Figure 14. Relative frequency distributions of wind direction at the 2 m for five stability regimes during the leaf-on (**a**) and leaf-off (**b**) seasons of 2012.

The below-canopy wind velocity also changed with subcanopy thermal gradient (Figure 15). During the night, the below-canopy valley wind velocity generally decreased with the below-canopy temperature inversion forming and increasing. Surprisingly again, the down-slope wind became weaker as inversion increasing. This indicated that the down-slope wind was suppressed by down-valley wind at situations with strong negative buoyancy force. During the daytime, however, the response of valley wind to temperature gradient was similar to that of the slope wind. However, the down-valley wind trended to be weakened by down-slope wind at conditions of strong negative buoyancy force in the leaf-on season.

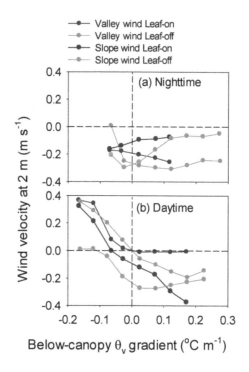

Figure 15. Change of mean velocity of valley and slope winds at the 2 m with the gradient of below-canopy virtual potential air temperature (θ_v) in the nighttime (**a**) and daytime (**b**) in 2012.

5. Conclusions

We examined the airflows above and below a temperate deciduous forest canopy in a valley at the Maoershan site of ChinaFLUX. In general, the above-canopy airflow was a typical thermally driven valley wind system, with the wind preferable up-valley in the daytime and down-valley in the nighttime (Table 2), which was consistent with the daytime temperature lapse and nocturnal inversion above the canopy (Figure 4a,b). Notably, the below-canopy airflow was often not aligned with that above canopy, especially in the leaf-on season (Table 2) and for above-canopy stable conditions (Figure 10). During the night in both leaf-on and leaf-off seasons, the below-canopy wind significantly sheared to the slope direction compared with the down-valley wind above canopy, which indicated a clear confluence of drainage flow onto the valley center. This drainage flow contradicted the prevailing temperature lapse and unstable conditions. During the daytime of the leaf-on season, the below-canopy layer prevailing direction was down-slope because of the frequent temperature inversion and isothermal conditions (Figure 4c), while during the leaf-off daytimes the below-canopy layer up-valley wind became more important due to the strong temperature lapse conditions (Figure 4d). The mean daytime positive and nighttime negative vertical velocity above the canopy during the leaf-on season (Figure 7a) indicated a flow convergence at night and a divergence in the daytime. This partly compensated by the below-canopy horizontal convergence in the daytime, but not in the nighttime. Overall, the below-canopy misalignment occurred at more than 70% of the time. The directional shear leaded to considerable lateral kinematic momentum fluxes (Figure 7), indicating a significantly interaction between slope and valley winds.

Above canopy, the nocturnal drainage maximum often occurred at the 42 m above the ground level or at the 24 m above the canopy (Figure 3), which was relative high compared with previous studies [19,28,38]. The drainage flow intensity (measured by the down-valley velocity at the 48 m and the velocity shear between 42 and 48 m) and probability were optimized when the above-canopy θ_v gradient was about 0.11 °C·m^{-1} during the leaf-on season and 0.08 °C·m^{-1} during the leaf-off season, respectively (Figures 9 and 11). In the daytime, the up-valley wind intensity and probability decreased as the θ_v gradient changed from lapse conditions to isothermal conditions. Below the canopy, the nocturnal drainage nearly persistently strengthened with above-canopy θ_v gradient increasing (Figures 9 and 11).

We found that the wind around the tower had a distinct 3-dimensional nature, which had significant implications for the mass flux measurements by EC method. First, inference about drainage flow based on above-canopy measurement alone may be not enough, since the down-slope wind prevails under the dense canopy even for unstable conditions, independent of the above-canopy wind (Figure 10a,c). Second, we found significant directional shear of horizontal wind, suggesting it might be inappropriate to simplify the horizontal advection to a 2-dimensional design such as at the Vielsalm site in Belgium [56], the Hainich site in Germany [53] and the Lageren site in Switzerland [46]. It may be more accurate to estimate the horizontal advection layer by layer, with the horizontal gradient of CO_2 concentration projected to the mean wind stream for each layer [57]. Third, the magnitude of vertical velocity at 36 m trended to vanish near ground, but the below-canopy vertical velocity could be opposite to that above canopy in the leaf-on mid-day (Figure 3a,c), which was due to the prevailing subcanopy down-slope wind but above-canopy up-valley wind (Table 2). This support the argument that the vertical advection based on one point vertical velocity measurement above canopy as a simplification are not generally applicable [58]. Thus, detailed profile of vertical velocity may be necessary to accurately estimate the total vertical advection by various layers with different vertical velocity and CO_2 concentration measurements [58]. However, the profile of vertical velocity is particularly difficult to measure because of various factors such as coordinate system selection and sonic anemometer bias [11,59,60]. And there is another challenge to "continuously" monitor the horizontal and vertical gradients of CO_2 concentration due to large temporal fluctuation, discrete temporal sampling and small magnitude gradients [6,60,61]. Fourth, the wind regime (and thus advection) at Maoershan site is different from those at the three sites in the Advection Experiment (ADVEX) [2,3], the Colorado site [5,14,18,57], the Morgan-Monroe State Forest site [10,62] and others [9,11,13,30], although there are some general patterns of thermally driven wind. These findings supported the advection regime is highly site-dependent [25]. The eddy flux measurements at valley sites are still a big challenge because of more complex wind regime than at slope sites. This inference was confirmed by the findings at two other valley sites [11,30]. Fifth, because direct advection measurements may be suffered from considerable errors [2], novel experiment design [13] and data processing strategies [11,38,63] may be helpful to gain robust annual fluxes of carbon and water vapor in complex terrain. On the other hand, comprehensive comparisons between the NEE by the EC technique with other independent measurements such as the ecosystem respiration by chamber-based up-scaling method and the biometric net ecosystem production [13,64,65] will be necessary to make a cross-validation of the carbon fluxes, which are ongoing at the Maoershan site.

In conclusion, we found a strong interaction between valley and slope winds at the Maoershan site. The relative high elevation of nocturnal drainage maximum (often 42 m) likely resulted from the unique terrain characteristics: the lower part of the sidewall of the valley and a long distance from the valley head. The above-canopy wind was a typical thermally driven valley wind system. However, below the canopy during the leaf-on season, down-slope wind predominated at night despite of a temperature lapse and unstable condition, whereas down-slope flow prevailed in the daytime due to frequent temperature inversion and isothermal conditions. The contrasting thermal gradients, stability regimes, and wind directions between below-canopy and above-canopy layers suggested a frequent decoupling of subcanopy layer. During the leaf-on season, the frequent directional shear and significant lateral kinematic momentum fluxes, above-canopy convergence in the nighttime and divergence in the daytime probably result in significant advective fluxes of trace gases and energy. A 3-dimensional design of wind vector and scalar concentration fields is essential to direct advection measurements in similar complex terrains.

Acknowledgments

The research was financially supported by the National Key Technology Research and Development Program of the Ministry of Science and Technology of China (No. 2011BAD37B01), the Program for Changjiang Scholars and Innovative Research Team in University (IRT1054) and the National Natural Science Funds (No. 30625010) to C. K. Wang. Special thanks go to two anonymous reviewers for their constructive comments. We also appreciate Jiquan Chen and Christian Feigenwinter for their constructive suggestions on the manuscript. The Maoershan Forest Ecosystem Research Station provided field logistic support.

Author Contributions

The paper was written by Xingchang Wang with a significant contribution by Chuankuan Wang and Qinglin Li. Chuankuan Wang and Xingchang Wang are responsible for the research design. Xingchang Wang conducted most of the field work.

Conflicts of Interest

The authors declare no conflict of interest.

References

1. Baldocchi, D. Measuring fluxes of trace gases and energy between ecosystems and the atmosphere—The state and future of the eddy covariance method. *Global. Change. Biol.* **2014**, *20*, 3600–3609.

2. Aubinet, M.; Feigenwinter, C.; Heinesch, B.; Bernhofer, C.; Canepa, E.; Lindroth, A.; Montagnani, L.; Rebmann, C.; Sedlak, P.; van Gorsel, E. Direct advection measurements do not help to solve the night-time CO_2 closure problem: Evidence from three different forests. *Agric. For. Meteorol.* **2010**, *150*, 655–664.

3. Feigenwinter, C.; Bernhofer, C.; Eichelmann, U.; Heinesch, B.; Hertel, M.; Janous, D.; Kolle, O.; Lagergren, F.; Lindroth, A.; Minerbi, S.; *et al.* Comparison of horizontal and vertical advective CO_2 fluxes at three forest sites. *Agric. For. Meteorol.* **2008**, *148*, 12–24.

4. Lee, X. On micrometeorological observations of surface-air exchange over tall vegetation. *Agric. For. Meteorol.* **1998**, *91*, 39–49.

5. Sun, J.; Burns, S.P.; Delany, A.C.; Oncley, S.P.; Turnipseed, A.A.; Stephens, B.B.; Lenschow, D.H.; LeMone, M.A.; Monson, R.K.; Anderson, D.E. CO_2 transport over complex terrain. *Agric. For. Meteorol.* **2007**, *145*, 1–21.

6. Van Gorsel, E.; Harman, I.N.; Finnigan, J.J.; Leuning, R. Decoupling of air flow above and in plant canopies and gravity waves affect micrometeorological estimates of net scalar exchange. *Agric. For. Meteorol.* **2011**, *151*, 927–933.

7. Staebler, R.M.; Fitzjarrald, D.R. Observing subcanopy CO_2 advection. *Agric. For. Meteorol.* **2004**, *122*, 139–156.

8. Tóta, J.; Fitzjarrald, D.R.; Staebler, R.M.; Sakai, R.K.; Moraes, O.M.; Acevedo, O.C.; Wofsy, S.C.; Manzi, A.O. Amazon rain forest subcanopy flow and the carbon budget: Santarém lba-eco site. *J. Geophys. Res.* **2008**, doi:10.1029/2007JG000597.

9. Alekseychik, P.; Mammarella, I.; Launiainen, S.; Rannik, Ü.; Vesala, T. Evolution of the nocturnal decoupled layer in a pine forest canopy. *Agric. For. Meteorol.* **2013**, *174–175*, 15–27.

10. Froelich, N.; Schmid, H. Flow divergence and density flows above and below a deciduous forest: Part II. Below-canopy thermotopographic flows. *Agric. For. Meteorol.* **2006**, *138*, 29–43.

11. Novick, K.; Brantley, S.; Miniat, C.F.; Walker, J.; Vose, J.M. Inferring the contribution of advection to total ecosystem scalar fluxes over a tall forest in complex terrain. *Agric. For. Meteorol.* **2014**, *185*, 1–13.

12. Belcher, S.E.; Harman, I.N.; Finnigan, J.J. The wind in the willows: Flows in forest canopies in complex terrain. *Annu. Rev. Fluid Mech.* **2012**, *44*, 479–504.

13. Thomas, C.K.; Martin, J.G.; Law, B.E.; Davis, K. Toward biologically meaningful net carbon exchange estimates for tall, dense canopies: Multi-level eddy covariance observations and canopy coupling regimes in a mature douglas-fir forest in oregon. *Agric. For. Meteorol.* **2013**, *173*, 14–27.

14. Yi, C.; Monson, R.K.; Zhai, Z.; Anderson, D.E.; Lamb, B.; Allwine, G.; Turnipseed, A.A.; Burns, S.P. Modeling and measuring the nocturnal drainage flow in a high-elevation, subalpine forest with complex terrain. *J. Geophys. Res.* **2005**, doi:10.1029/2005JD006282.

15. Zardi, D.; Whiteman, C.D. Diurnal mountain wind systems. In *Mountain Weather Research and Forecasting: Recent Progress and Current Challenges*; Chow, F., de Wekker, S., Snyder, B., Eds.; Springer: Berlin, Germany, 2013; pp 35–119.

16. Grisogono, B.; Axelsen, S.L. A note on the pure katabatic wind maximum over gentle slopes. *Bound.-Layer Meteorol.* **2012**, *145*, 527–538.

17. Chen, H.; Yi, C. Optimal control of katabatic flows within canopies. *Q. J. R. Meteor. Soc.* **2012**, *138*, 1676–1680.

18. Burns, S.P.; Sun, J.; Lenschow, D.H.; Oncley, S.P.; Stephens, B.B.; Yi, C.; Anderson, D.E.; Hu, J.; Monson, R.K. Atmospheric stability effects on wind fields and scalar mixing within and just above a subalpine forest in sloping terrain. *Bound.-Layer Meteorol.* **2011**, *138*, 231–262.

19. Aubinet, M.; Berbigier, P.; Bernhofer, C.; Cescatti, A.; Feigenwinter, C.; Granier, A.; Grünwald, T.; Havrankova, K.; Heinesch, B.; Longdoz, B. Comparing CO_2 storage and advection conditions at night at different carboeuroflux sites. *Bound.-Layer Meteorol.* **2005**, *116*, 63–93.

20. Whiteman, C.D.; Zhong, S. Downslope flows on a low-angle slope and their interactions with valley inversions. Part I: Observations. *J. Appl. Meteorol. Climitol.* **2008**, *47*, 2023–2038.

21. Trachte, K.; Nauss, T.; Bendix, J. The impact of different terrain configurations on the formation and dynamics of katabatic flows: Idealised case studies. *Bound.-Layer Meteorol.* **2010**, *134*, 307–325.

22. van Gorsel, E.; Christen, A.; Feigenwinter, C.; Parlow, E.; Vogt, R. Daytime turbulence statistics above a steep forested slope. *Bound.-Layer Meteorol.* **2003**, *109*, 311–329.

23. Mahrt, L.; Richardson, S.; Seaman, N.; Stauffer, D. Non-stationary drainage flows and motions in the cold pool. *Tellus A* **2010**, *62*, 698–705.

24. Belcher, S.; Finnigan, J.; Harman, I. Flows through forest canopies in complex terrain. *Ecol. Appl.* **2008**, *18*, 1436–1453.

25. Aubinet, M. Eddy covariance CO_2 flux measurements in nocturnal conditions: An analysis of the problem. *Ecol. Appl.* **2008**, *18*, 1368–1378.

26. Kiefer, M.T.; Zhong, S. The effect of sidewall forest canopies on the formation of cold-air pools: A numerical study. *J. Geophys. Res.* **2013**, *118*, 5965–5978.

27. Froelich, N.; Grimmond, C.; Schmid, H. Nocturnal cooling below a forest canopy: Model and evaluation. *Agric. For. Meteorol.* **2011**, *151*, 957–968.

28. Mahrt, L. Stratified atmospheric boundary layers. *Bound.-Layer Meteorol.* **1999**, *90*, 375–396.

29. Pypker, T.; Unsworth, M.; Lamb, B.; Allwine, E.; Edburg, S.; Sulzman, E.; Mix, A.; Bond, B. Cold air drainage in a forested valley: Investigating the feasibility of monitoring ecosystem metabolism. *Agric. For. Meteorol.* **2007**, *145*, 149–166.

30. Tóta, J.; Roy Fitzjarrald, D.; da Silva Dias, M.A. Amazon rainforest exchange of carbon and subcanopy air flow: Manaus lba site—A complex terrain condition. *Sci. World J.* **2012**, doi:10.1100/2012/165067.

31. Komatsu, H.; Yoshida, N.; Takizawa, H.; Kosaka, I.; Tantasirin, C.; Suzuki, M. Seasonal trend in the occurrence of nocturnal drainage flow on a forested slope under a tropical monsoon climate. *Bound.-Layer Meteorol.* **2003**, *106*, 573–592.

32. Pypker, T.G.; Unsworth, M.H.; Mix, A.C.; Rugh, W.; Ocheltree, T.; Alstad, K.; Bond, B.J. Using nocturnal cold air drainage flow to monitor ecosystem processes in complex terrain. *Ecol. Appl.* **2007**, *17*, 702–714.

33. Rotach, M.; Andretta, M.; Calanca, P.; Weigel, A.; Weiss, A. Boundary layer characteristics and turbulent exchange mechanisms in highly complex terrain. *Acta Geophys.* **2008**, *56*, 194–219.

34. Jiao, Z.; Wang, C.-K.; Wang, X.-C. Spatio-temporal variations of CO_2 concentration within the canopy in a temperate deciduous forest, northeast China. *Chin. J. Plant Ecol.* **2011**, *35*, 512–522.

35. Mahrt, L. Momentum balance of gravity flows. *J. Atmos. Sci.* **1982**, *39*, 2701–2711.

36. Mahrt, L.; Vickers, D.; Nakamura, R.; Soler, M.R.; Sun, J.; Burns, S.; Lenschow, D.H. Shallow drainage flows. *Bound.-Layer Meteorol.* **2001**, *101*, 243–260.

37. Rotach, M.W.; Zardi, D. On the boundary-layer structure over highly complex terrain: Key findings from map. *Q. J. R. Meteor. Soc.* **2007**, *133*, 937–948.

38. Rebmann, C.; Zeri, M.; Lasslop, G.; Mund, M.; Kolle, O.F.; Schulze, E.-D.; Feigenwinter, C. Treatment and assessment of the CO_2-exchange at a complex forest site in thuringia, germany. *Agric. For. Meteorol.* **2010**, *150*, 684–691.

39. Staebler, R.M.; Fitzjarrald, D.R. Measuring canopy structure and the kinematics of subcanopy flows in two forests. *J. Appl. Meteorol.* **2005**, *44*, 1161–1179.

40. Wang, C.; Han, Y.; Chen, J.; Wang, X.; Zhang, Q.; Bond-Lamberty, B. Seasonality of soil CO_2 efflux in a temperate forest: Biophysical effects of snowpack and spring freeze–thaw cycles. *Agric. For. Meteorol.* **2013**, *177*, 83–92.

41. Wilczak, J.M.; Oncley, S.P.; Stage, S.A. Sonic anemometer tilt correction algorithms. *Bound.-Layer Meteorol.* **2001**, *99*, 127–150.

42. Yi, C. Momentum transfer within canopies. *J. Appl. Meteorol. Climitol.* **2008**, *47*, 262–275.

43. Mahrt, L.; Lee, X.; Black, A.; Neumann, H.; Staebler, R. Nocturnal mixing in a forest subcanopy. *Agric. For. Meteorol.* **2000**, *101*, 67–78.

44. Mahrt, L.; Vickers, D.; Sun, J.; Jensen, N.; Jørgensen, H.; Pardyjak, E.; Fernando, H. Determination of the surface drag coefficient. *Bound.-Layer Meteorol.* **2001**, *99*, 249–276.

45. Weber, R. Remarks on the definition and estimation of friction velocity. *Bound.-Layer Meteorol.* **1999**, *93*, 197–209.

46. Dupont, S.; Patton, E.G. Influence of stability and seasonal canopy changes on micrometeorology within and above an orchard canopy: The chats experiment. *Agric. For. Meteorol.* **2012**, *157*, 11–29.

47. Yao, Y.; Zhang, Y.; Liang, N.; Tan, Z.; Yu, G.; Sha, L.; Song, Q. Pooling of CO_2 within a small valley in a tropical seasonal rain forest. *J. Forest. Res.* **2012**, *17*, 241–252.

48. Monti, P.; Fernando, H.; Princevac, M.; Chan, W.; Kowalewski, T.; Pardyjak, E. Observations of flow and turbulence in the nocturnal boundary layer over a slope. *J. Atmos. Sci.* **2002**, *59*, 2513–2534.

49. Burns, P.; Chemel, C. Evolution of cold-air-pooling processes in complex terrain. *Bound.-Layer Meteorol.* **2014**, *150*, 423–447.

50. Thomas, C.K. Variability of sub-canopy flow, temperature, and horizontal advection in moderately complex terrain. *Bound.-Layer Meteorol.* **2011**, *139*, 61–81.

51. Wu, J.; Guan, D.; Yuan, F.; Yang, H.; Wang, A.; Jin, C. Evolution of atmospheric carbon dioxide concentration at different temporal scales recorded in a tall forest. *Atmos. Environ.* **2012**, *61*, 9–14.

52. Aubinet, M.; Heinesch, B.; Yernaux, M. Horizontal and vertical CO_2 advection in a sloping forest. *Bound.-Layer Meteorol.* **2003**, *108*, 397–417.

53. Kruijt, B.; Malhi, Y.; Lloyd, J.; Norbre, A.; Miranda, A.; Pereira, M.; Culf, A.; Grace, J. Turbulence statistics above and within two amazon rain forest canopies. *Bound.-Layer Meteorol.* **2000**, *94*, 297–331.

54. Sedlák, P.; Aubinet, M.; Heinesch, B.; Janouš, D.; Pavelka, M.; Potužníková, K.; Yernaux, M. Night-time airflow in a forest canopy near a mountain crest. *Agric. For. Meteorol.* **2010**, *150*, 736–744.

55. Bernardes, M.; Dias, N.L. The alignment of the mean wind and stress vectors in the unstable surface layer. *Bound.-Layer Meteorol.* **2010**, *134*, 41–59.

56. Su, H.-B.; Schmid, H.; Vogel, C.; Curtis, P. Effects of canopy morphology and thermal stability on mean flow and turbulence statistics observed inside a mixed hardwood forest. *Agric. For. Meteorol.* **2008**, *148*, 862–882.

57. Yi, C.; Anderson, D.E.; Turnipseed, A.A.; Burns, S.P.; Sparks, J.P.; Stannard, D.I.; Monson, R.K. The contribution of advective fluxes to net ecosystem exchange in a high-elevation, subalpine forest. *Ecol. Appl.* **2008**, *18*, 1379–1390.

58. Feigenwinter, C.; Montagnani, L.; Aubinet, M. Plot-scale vertical and horizontal transport of CO_2 modified by a persistent slope wind system in and above an alpine forest. *Agric. For. Meteorol.* **2010**, *150*, 665–673.

59. Queck, R.; Bernhofer, C. Constructing wind profiles in forests from limited measurements of wind and vegetation structure. *Agric. For. Meteorol.* **2010**, *150*, 724–735.

60. Siebicke, L.; Hunner, M.; Foken, T. Aspects of CO_2 advection measurements. *Theor. Appl. Climatol.* **2012**, *109*, 109–131.

61. Marcolla, B.; Cobbe, I.; Minerbi, S.; Montagnani, L.; Cescatti, A. Methods and uncertainties in the experimental assessment of horizontal advection. *Agric. For. Meteorol.* **2014**, *198–199*, 62–71.

62. Froelich, N.; Schmid, H.; Grimmond, C.; Su, H.-B.; Oliphant, A. Flow divergence and density flows above and below a deciduous forest: Part I. Non-zero mean vertical wind above canopy. *Agric. For. Meteorol.* **2005**, *133*, 140–152.

63. van Gorsel, E.; Delpierre, N.; Leuning, R.; Black, A.; Munger, J.W.; Wofsy, S.; Aubinet, M.; Feigenwinter, C.; Beringer, J.; Bonal, D. Estimating nocturnal ecosystem respiration from the vertical turbulent flux and change in storage of CO_2. *Agric. For. Meteorol.* **2009**, *149*, 1919–1930.

64. Gielen, B.; de Vos, B.; Campioli, M.; Neirynck, J.; Papale, D.; Verstraeten, A.; Ceulemans, R.; Janssens, I. Biometric and eddy covariance-based assessment of decadal carbon sequestration of a temperate scots pine forest. *Agric. For. Meteorol.* **2013**, *174*, 135–143.

65. Wu, J.; Larsen, K.; van der Linden, L.; Beier, C.; Pilegaard, K.; Ibrom, A. Synthesis on the carbon budget and cycling in a danish, temperate deciduous forest. *Agric. For. Meteorol.* **2013**, *181*, 94–107.

7

PM$_{2.5}$ Chemical Compositions and Aerosol Optical Properties in Beijing during the Late Fall

Huanbo Wang [1], Xinghua Li [2], Guangming Shi [1], Junji Cao [3], Chengcai Li [4], Fumo Yang [1,*], Yongliang Ma [5] and Kebin He [5]

[1] Key Laboratory of Reservoir Aquatic Environment of CAS, Chongqing Institute of Green and Intelligent Technology, Chinese Academy of Sciences, Chongqing 400714, China;
E-Mails: hbwang@cigit.ac.cn (H.W.); shigm@cigit.ac.cn (G.S.)

[2] School of Chemistry and Environment, Beihang University, Beijing 100191, China;
E-Mail: lixinghua@buaa.edu.cn

[3] Key Lab of Aerosol Chemistry & Physics, Institute of Earth Environment, Chinese Academy of Sciences, Xi'an 710075, China; E-Mail: cao@loess.llqg.ac.cn

[4] School of Physics, Peking University, Beijing 100871, China; E-Mail: ccli@pku.edu.cn

[5] State Environmental Protection Key Laboratory of Sources and Control of Air Pollution Complex, School of Environment, Tsinghua University, Beijing 100084, China;
E-Mails: liang@tsinghua.edu.cn (Y.M.); hekb@tsinghua.edu.cn (K.H.)

* Author to whom correspondence should be addressed; E-Mail: fmyang@cigit.ac.cn

Academic Editors: Ru-Jin Huang and Robert W. Talbot

Abstract: Daily PM$_{2.5}$ mass concentrations and chemical compositions together with the aerosol optical properties were measured from 8–28 November 2011 in Beijing. PM$_{2.5}$ mass concentration varied from 15.6–237.5 $\mu g \cdot m^{-3}$ and showed a mean value of 111.2 ± 73.4 $\mu g \cdot m^{-3}$. Organic matter, NH$_4$NO$_3$ and (NH$_4$)$_2$SO$_4$ were the major constituents of PM$_{2.5}$, accounting for 39.4%, 15.4%, and 14.9% of the total mass, respectively, while fine soil, chloride salt, and elemental carbon together accounted for 27.7%. Daily scattering and absorption coefficients (σ_{sc} and σ_{ap}) were in the range of 31.1–667 Mm^{-1} and 8.24–158.0 Mm^{-1}, with mean values of 270 ± 200 Mm^{-1} and 74.3 ± 43.4 Mm^{-1}. Significant increases in σ_{sc} and σ_{ap} were observed during the pollution accumulation episodes. The revised IMPROVE algorithm was applied to estimate the extinction coefficient (b$_{ext}$). On average, organic matter was the largest contributor, accounting for

44.6% of b_{ext}, while $(NH_4)_2SO_4$, NH_4NO_3, elemental carbon, and fine soil accounted for 16.3% 18.0%, 18.6%, and 2.34% of b_{ext}, respectively. Nevertheless, the contributions of $(NH_4)_2SO_4$ and NH_4NO_3 were significantly higher during the heavy pollution periods than those on clean days. Typical pollution episodes were also explored, and it has been characterized that secondary formation of inorganic compounds is more important than carbonaceous pollution for visibility impairment in Beijing.

Keywords: $PM_{2.5}$; visibility; aerosol optical properties; chemical composition

1. Introduction

The atmospheric visibility in China has been deteriorating with economic growth during the past 40 years [1,2]. Visibility impairment is resulted from light scattering and absorption by atmospheric particles and gases, especially from the scattering by the particles of similar size range as the wavelength range of visible light. Numerous studies have indicated that the fine particles caused most of the visibility impairment, while the influence of gas and coarse particles on visibility degradation was commonly weak [3,4]. Moreover, meteorological parameters, such as wind, rain, and temperature, especially the relative humidity, have their contributions as well [5]. Additionally, atmospheric particles have significant impacts on climate change, which is one of the greatest sources of uncertainty in estimating the direct radiative forcing [6,7]. Generally, inorganic and organic aerosols have a cooling effect on climate by scattering light, while black carbon (BC) has a warming effect by absorbing light.

Previous studies usually determined the chemical compositions and optical properties of atmospheric aerosols separately [8–10]. Yang *et al.* [11] compared the characteristics of $PM_{2.5}$ in representative megacities of China. Results showed that five major species including organic carbon (OC), elemental carbon (EC), SO_4^{2-}, NO_3^-, and NH_4^+ amounted to 54%–59% of $PM_{2.5}$ mass in Beijing, Chongqing and Guangzhou, and the percentages of total carbon and secondary inorganic ions were very close, implying that both primary and secondary particles had a significant contribution to the $PM_{2.5}$ mass. Recently, Zhang *et al.* [12] indicated that secondary inorganic aerosols, mineral dust and organic matter (OM) each accounted for about 20% of $PM_{2.5}$ in Beijing, respectively, suggesting both primary and secondary components of $PM_{2.5}$ in Beijing were equally important.

The parameters of light extinction (b_{ext}) can be measured directly using optical instruments such as an integrating nephelometer for the light scattering coefficient (σ_{sc}), or an aethalometer for the absorption coefficients (σ_{ap}). Optical properties of $PM_{2.5}$ have been conducted in Beijing, Shanghai, and Guangzhou [13–17]. In recent years, a few studies have focused on the relationship between the chemical compositions and optical properties of aerosols [18,19]. Results showed that NH_4^+, SO_4^{2-}, NO_3^-, and OC are the main contributors to aerosol scattering, and the light absorption coefficients had strong linear correlations with EC in Shanghai. Furthermore, as an alternative method, the Interagency Monitoring of Protected Visual Environments (IMPROVE) formula could be used to estimate b_{ext} based on the chemical compositions of particulate matter (PM). The original and revised IMPROVE algorithms for estimating b_{ext} were developed from the particle data at 21 rural/remote sites with low scattering coefficients. Furthermore, the IMPROVE formula usually assumes externally mixed status of PM and fixed mass extinction efficiency for each species [20]. However, the actual mass absorption efficiency (MAE) and mass scattering efficiency (MSE) were not constant due to large temporal and spatial variations of chemical compositions of PM. Thus, it is necessary to evaluate the applicability of the IMPROVE formula to the calculation of b_{ext} in more polluted urban Beijing in China.

In the present study, aerosol optical properties including σ_{sc} and σ_{ap}, as well as the chemical compositions of PM$_{2.5}$ were measured in Beijing during the late fall of 2011. The applicability of the IMPROVE formula to the calculation of b_{ext} was then evaluated, and the contribution of PM$_{2.5}$ chemical compositions to visibility impairment was discussed. In addition, the formation mechanisms of typical pollution episodes during the late fall were also explored.

2. Experimental Section

2.1. Sampling

Samples were collected from 8–28 November 2011 at the campus of Tsinghua University (39°98′N, 116°32′E) in urban Beijing, about 600 m north of the Fourth Ring Road. The campus is mainly surrounded by residential areas without significant factory emissions. Beijing is connected to the industrialized cities of the Great North China Plain in the South, and surrounded by the Yanshan Mountains in the west, north, and northeast.

Daily 23 h integrated PM$_{2.5}$ samples were collected using a five-channel Spiral Ambient Speciation Sampler (SASS, MetOne Inc., Grants Pass, OR, USA) with a flow rate of 6.7 L·min^{-1}. The first channel was used for PM$_{2.5}$ mass and elemental analysis with a 47 mm Teflon filter. The second channel collected the particles for the analysis of water-soluble inorganic ions with a 47 mm Teflon filter. The third channel was used to collect PM$_{2.5}$ on quartz filters for organic and elemental carbon analysis.

2.2. Gravimetric and Chemical Analysis

The PM$_{2.5}$ mass concentrations were determined using an electronic balance with a detection limit of 1 μg (Sartorius, Göttingen, Germany) after stabilizing at a constant temperature (22 ± 5 °C) and relative humidity (40% ± 5%) for 24 h.

Four anions (SO$_4^{2-}$, NO$_3^-$, Cl$^-$, and F$^-$) and five cations (Na$^+$, NH$_4^+$, K$^+$, Mg^{2+}, and Ca^{2+}) were determined in aqueous extracts of the filters by Ion chromatography (ICS-1000 and ICS-2000 for anion and cation, respectively, Dionex, Sunnyvale, CA, USA). To extract the water-soluble ions from the Teflon filters, each sample was extracted twice with 7.5 ml Milli-Q water via an ultrasonic bath for 20 min, and then filtered through a 0.45 μm PTFE syringe filter and stored in a refrigerator at 4 °C until analysis.

A 0.5 cm^2 punch from each quartz filter was analyzed for OC and EC using a DRI Model 2001 Thermal/Optical Carbon Analyzer (Atmoslytic Inc., Calabasas, CA, USA), following the IMPROVE thermal optical reflectance (TOR) protocol [21].

Crustal elements including Al, Si, Ca, Fe, and Ti were analyzed by Energy Dispersive X-ray fluorescence spectrometry (Epsilon 5 ED-XRF, PANalytical Company, Almelo, The Netherlands) on Teflon filters. Quality assurance/Quality Control (QA/QC) procedures of the XRF analysis procedure were described by Xu *et al.* [22].

2.3. Quality Control

The fresh quartz filters were pre-heated at 450 °C in a muffle furnace for 6 h to remove any volatile components before sampling. Furthermore, after collection, the samples were sealed in clean plastic bags, and were stored in a freezer at −18 °C before chemical analysis to minimize the evaporation of volatile components. Before and after sampling, the Teflon filters in the first channel were weighed after being equilibrated for 24 h. The artifacts during the sampling and analysis were estimated by a field blank filter.

2.4. Measurements of Aerosol Optical and Meteorological Parameters

BC mass concentration, σ_{sc} and meteorological data were measured on the roof of the Physics Building about 30 m above the ground in Peking University, which is about 1 km away from the sampling sites of Tsinghua University. An automatic weather station (Vaisala Ltd., Helsinki, Finland) was used to record wind speed (WS), wind direction, relative humidity (RH), temperature (Temp), and visibility (VR).

σ_{sc} was monitored using a single wavelength (525 nm) integrating Nephelometer (M9003, Ecotech, Melbourne, VIC, Australia). This instrument drew ambient air through a heated inlet tube to maintain RH in the Nephelometer chamber below 60%. The scattering intensity over angles from 7° to 170° was measured and integrated to yield the scattering coefficient. Zero calibration was performed every two days with particle-free air to subtract the Rayleigh scattering, while span calibration was carried out every month using R-134 gas.

BC mass concentration was measured with an Aethalometer (AE-16, Magee Scientific, Berkeley, CA, USA). The principle of this instrument to calculate the BC concentration is based on the attenuation of an incident beam at a wavelength of 800 nm caused by the particles loaded in the quartz filter. No size-selective inlet was used for both the nephelometer and aethalometer. Considering the negligible contribution of coarse particles to light extinction, the measured σ_{sc} can be approximately attributed to the $PM_{2.5}$.

2.5. Data Analysis

2.5.1. Reconstruction of $PM_{2.5}$ Mass

$PM_{2.5}$ components can be grouped as follows: secondary inorganic aerosols (SNA), OM, EC, fine soil (FS), and chloride salt (CS). SNA is the sum of SO_4^{2-}, NO_3^-, and NH_4^+, and OM is derived from multiplying OC concentrations by a factor of 1.6 to account for unmeasured atoms according to Xing et al. [23], which demonstrated that the calculated OM/OC mass ratio in summer was relatively high (1.75 ± 0.13) and in winter was lower (1.59 ± 0.18) in $PM_{2.5}$ collected from 14 Chinese cities. Beijing is far from the coastal oceans, and sea salt is not transported to Beijing, thus it has a minor contribution to $PM_{2.5}$ in Beijing. The CS was considered instead of sea salt, and estimated by summing concentrations of Cl^-, K^+, and Na^+ according to Zhang et al. [12].

The concentrations of FS are often estimated by assuming the oxides of the elements mainly associated with soil (Al_2O_3, SiO_2, K_2O, CaO, FeO, Fe_2O_3, and TiO_2), which is calculated as follows [24]:

$$[FS] = 2.20[Al] + 2.49[Si] + 1.63[Ca] + 2.42[Fe] + 1.94[Ti] \tag{1}$$

2.5.2. Reconstruction of the Light Extinction Coefficient

According to the revised IMPROVE algorithm, the reconstructed b_{ext} is shown from the following equation assuming an externally mixed aerosol [20]:

$$\begin{aligned}
b_{ext} &= b_{sp} + b_{ap} + b_{ag} + b_{sg} \\
&\approx 2.2 \times f_s(RH) \times [Small(NH_4)_2SO_4] + 4.8 \times f_L(RH) \times [Large(NH_4)_2SO_4] \\
&\quad + 2.4 \times f_s(RH) \times [SmallNH_4NO_3] + 5.1 \times f_L(RH) \times [Large NH_4NO_3] \\
&\quad + 2.8 \times [SmallOM] + 6.1 \times [Large OM] \\
&\quad + 10 \times [EC] + 1 \times [Fine\ Soil] + 1.7 \times f_{ss}(RH) \times [Sea\ Salt] + 0.6 \times [PM_{2.5\sim10}] \\
&\quad + 0.33 \times [NO_2(ppb)] + Rayleigh\ Scattering
\end{aligned} \tag{2}$$

The algorithm divides the concentrations of SO_4^{2-}, NO_3^-, and OM into small and large-sized fractions. The size modes are described by log-normal mass size distributions with geometric mean diameter and geometric standard deviations. The fraction of a component in the large- or small-sized mode was estimated by an empirical approach [20]. The apportionment of the total concentrations of $(NH_4)_2SO_4$ into the concentrations of the small and large size fraction in PM2.5 is accomplished using the following equations:

$$[Large(NH_4)_2SO_4] = \frac{[Total(NH_4)_2SO_4]}{20\mu g.m^{-3}} \times [Total(NH_4)_2SO_4],\ for[Total(NH_4)_2SO_4] < 20\mu g.m^{-3} \tag{3}$$

$$[Large(NH_4)_2SO_4] = [Total(NH_4)_2SO_4],\ for[Total(NH_4)_2SO_4] > 20\mu g.m^{-3} \tag{4}$$

$$[Small(NH_4)_2SO_4] = [Total(NH_4)_2SO_4] - [Large(NH_4)_2SO_4] \tag{5}$$

Similar equations are used to apportion total NH_4NO_3 and total OM concentrations into small and large size fractions. The water growth adjustment term $f_s(RH)$, $f_L(RH)$ for small and large size distribution of $(NH_4)_2SO_4$ and NH_4NO_3, and $f_{ss}(RH)$ for sea salt are used according to the water growth curves provided by Pitchford et al. [20].

According to the revised IMPROVE method, SO_4^{2-} and NO_3^- are assumed to be fully neutralized by NH_4^+ in the forms of $(NH_4)_2SO_4$ and NH_4NO_3, respectively. Therefore, $(NH_4)_2SO_4$ mass is estimated by the SO_4^{2-} mass multiplied by a factor of 1.38, and the NH_4NO_3 mass is estimated by the NO_3^- mass multiplied by a factor of 1.29.

In order to compare with the reconstructed b_{ext} calculated using Mie theory at 550 nm, σ_{sc} measured at 525 nm with the integrating nephelometer should be converted to that at 550 nm according to the method by Jung et al. [16].

$$\sigma_{sc}(550nm) = \sigma_{sc}(525nm) * \left(\frac{\lambda(550nm)}{\lambda(525nm)} \right)^{-\alpha} \tag{6}$$

where α is the scattering Angström exponent, an average α value of 1.18 determined in Beijing during the summer of 2012 by Tian et al. [18] was used in the present study.

σ_{ab} at 550 nm was calculated based on BC concentration following the equation:

$$\sigma_{ab} = K \times [BC] \tag{7}$$

where K is the conversion factor, which was set to 8.1 $m^2 \cdot g^{-1}$ in this study according to a previous study [25].

The single scattering albedo (SSA) is defined as the ratio of the aerosol scattering coefficient to the extinction coefficient at a known wavelength, as derived from the formula:

$$SSA = \frac{\sigma_{sc}}{\sigma_{sc} + \sigma_{ab}} \tag{8}$$

3. Results and Discussion

3.1. PM2.5 Chemical Compositions

The time series of daily PM2.5 mass concentrations and the meteorological parameters, including RH, temperature, and WS are shown in Figure 1. The mass concentrations of PM2.5 ranged from 15.6–237.5 $\mu g \cdot m^{-3}$ and averaged 111.2 ± 73.4 $\mu g \cdot m^{-3}$. Compared with other studies conducted in urban Beijing, the average PM2.5 concentration in this study was lower than that measured during autumn of 2006 (194.2 $\mu g \cdot m^{-3}$) [26] and 2009 (135 $\mu g \cdot m^{-3}$) [12], while comparable with that observed during the same season in 2005 (115.0 $\mu g \cdot m^{-3}$) and 2012 (106.9 $\mu g \cdot m^{-3}$) [27]. There were sixty percent of days with daily PM2.5 mass concentration exceeding the China Ambient Air Quality Standards (75 $\mu g \cdot m^{-3}$). The highest PM2.5 mass concentration occurred on 26 November, which was associated with the high relative humidity and low wind speed.

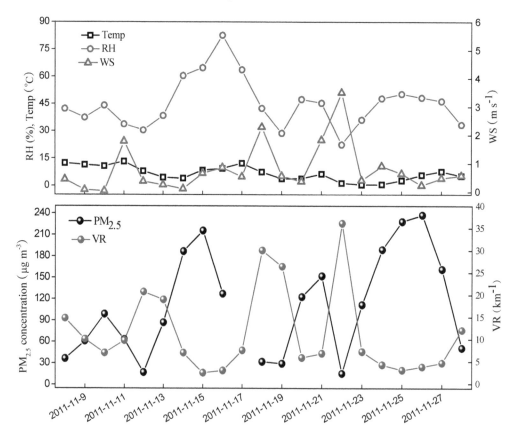

Figure 1. Daily variations of PM2.5 mass concentration and meteorological parameters.

The temporal variations of nine water-soluble inorganic ions (WSIIs) are presented in Figure 2. The average concentration of total nine ions was 46.9 ± 33.8 $\mu g \cdot m^{-3}$, accounting for 41.5% of $PM_{2.5}$ mass concentration. NO_3^- was the most abundant species in WSIIs with an average concentration of 14.7 ± 11.2 $\mu g \cdot m^{-3}$, followed by SO_4^{2-} (12.2 ± 9.63 $\mu g \cdot m^{-3}$), NH_4^+ (9.13 ± 7.26 $\mu g \cdot m^{-3}$), and Cl^- (6.62 ± 4.62 $\mu g \cdot m^{-3}$), accounting for 28.9%, 25.6%, 17.7%, and 14.6% of WSIIs, respectively. The rest of K^+ (1.66 ± 1.41 $\mu g \cdot m^{-3}$), Na^+ (0.95 ± 0.52 $\mu g \cdot m^{-3}$), Ca^{2+} (0.81 ± 0.32 $\mu g \cdot m^{-3}$), F^- (0.61 ± 0.34 $\mu g \cdot m^{-3}$), and Mg^{2+} (0.25 ± 0.18 $\mu g \cdot m^{-3}$) each had a minor contribution to the WSIIs, totally accounting for 13.2% of WSIIs. SNA typically constituted 33.5%–87.1% of the total WSIIs and 15.3%–46.0% of $PM_{2.5}$, respectively.

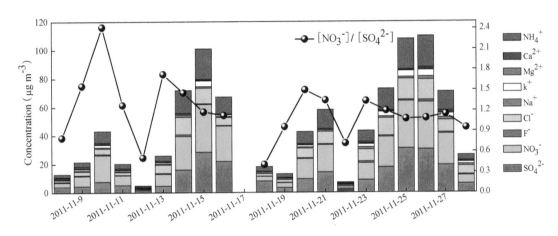

Figure 2. Daily variations of water-soluble ions and $[NO_3^-]/[SO_4^{2-}]$ ratios.

NO_3^- and SO_4^{2-} are mainly formed by atmospheric reactions of precursor gases such as NO_x and SO_2. Generally, SO_2 emits from coal combustion, while NO_x is the result of any type of combustion such as coal-fired power plants and automobiles. The mass ratio $[NO_3^-]/[SO_4^{2-}]$ has been used to identify the influence of the stationary and mobile sources of sulfur and nitrogen [28]. The $[NO_3^-]/[SO_4^{2-}]$ ratio ranged from 0.41–2.42, with an average value of 1.19. It was higher than that of values (around 0.68) measured in Beijing from 2001–2006 [26,29–31], but rather more comparable to those observed in recent years [12]. As illustrated in Figure 2, the ratio was usually lower during weekend days (12, 18–19 November) than on workdays, indicating that the higher $[NO_3^-]/[SO_4^{2-}]$ ratio in the present study was probably associated with the rapid increase of motor vehicles in recent years. According to the statistics from the China Vehicle Emission Control Annual Report in 2013, the amount of vehicles reached 5 million in Beijing by 2012, which was an increase of about three times compared with the amount of vehicles in 2001 [32].

Previous study showed that Cl^- might be derived from coal combustion when the Cl^-/Na^+ equivalent concentration ratios were larger than the mean ratio (1.17) for sea water. The ratios of Cl^-/Na^+ were in the range of 1.6–11.6 with a mean value of 6.63 during the study period, implying that Cl^- may be originated from coal combustion rather than sea spray [12].

The equivalent molar ratio of total cations to total anions (CE/AE) ranged from 0.71–1.40, with an average value of 0.95 ± 0.14 during the study period. Figure 3 illustrates the scatter plots of the sum of cations *versus* anions. Results showed that the slope was slightly lower than 1, implying that the fine particles collected in the study period were weakly acidic. Moreover, the ratios of $[NH_4^+]/[SO_4^{2-} + NO_3^-]$

were close to 1, demonstrating that SO_4^{2-} and NO_3^- were fully neutralized by NH_3. Therefore, the dominant chemical form of SO_4^{2-} was $(NH_4)_2SO_4$ rather than NH_4HSO_4, which can be estimated by the SO_4^{2-} mass concentration multiplied by a factor of 1.38, while NO_3^- existed as NH_4NO_3, and can be estimated by the NO_3^- mass concentration multiplied by a factor of 1.29.

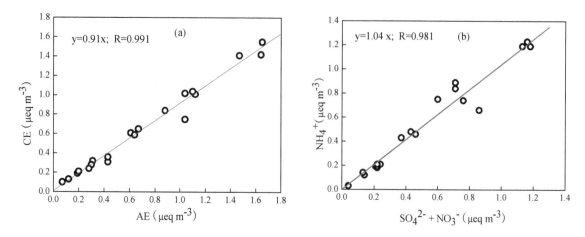

Figure 3. Relationships of equivalent concentrations of cations *versus* anions (**a**) and $[NH_4^+]$ *versus* $[SO_4^{2-} + NO_3^-]$ (**b**).

As illustrated in Figure 4, OC varied from 2.1 to 64.3 $\mu g \cdot m^{-3}$, averaging 27.5 ± 19.9 $\mu g \cdot m^{-3}$, while EC ranged from 0.86 to 19.6 $\mu g \cdot m^{-3}$, averaging 9.62 ± 6.24 $\mu g \cdot m^{-3}$. The contribution of OC and EC to $PM_{2.5}$ were 24.5% and 8.96%, respectively. Such levels of OC and EC were close to those observed in the same season in recent years [10,26,30,33], whereas lower than those measured ten years ago [29,34].

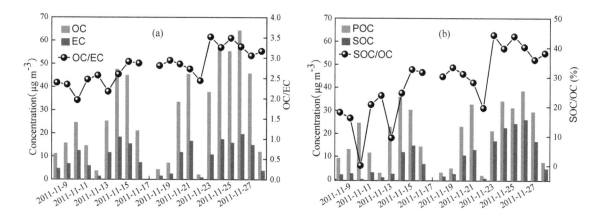

Figure 4. Variations of OC, EC, and OC/EC (**a**) as well as primary organic carbon (POC), secondary organic carbon (SOC), and SOC/OC (**b**).

The relationships between OC and EC can be used to identify the origins of carbonaceous particles [35,36]. As shown in Figure 5, strong correlations between OC and EC were observed with a correlation coefficient of 0.97, indicating that OC and EC were likely derived from the same major primary sources during the campaigns. On the other hand, the OC/EC ratios did not vary distinctly during the study period, especially during the space heating days. The ratios ranged from 1.96–3.52, averaging 2.79, and were very close to the value of 2.7 from coal combustion suggested by

Watson *et al.* [37]. This pointed to the fact that OC and EC likely originated mainly from coal combustions. Furthermore, the mean OC/EC ratio was higher than 2, indicating that SOC might be present during the study period [38].

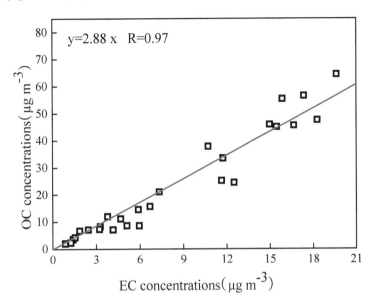

Figure 5. Relationships of OC and EC concentrations.

The method of EC-tracer has been widely used to estimate the SOC concentration since it was first introduced by Castro *et al.* [9,39]. This approach suggested that samples having the lowest OC/EC ratio contained almost exclusively POC. Then, the concentration of SOC can be estimated by the following formula:

$$POC = EC \times (OC/EC)_{min} \tag{9}$$

$$SOC = OC - POC \tag{10}$$

where $(OC/EC)_{min}$ was the value of the lowest OC/EC ratio. Based on the $(OC/EC)_{min}$ of 1.63, the SOC concentrations varied from 0.02–25.9 $\mu g \cdot m^{-3}$ with an average value of 9.03 $\mu g \cdot m^{-3}$. As illustrated in Figure 4, it is interesting to note that the concentrations of SOC were still high with low temperature during the study period except on 10, 12 and 22 November. This may be caused by the combination of the high precursor emission due to the largely increased coal combustion for residential heating and low wind speed (averaging 0.79 $m \cdot s^{-1}$), which was favorable for the pollutants accumulation and formation of secondary organic aerosol. The low temperature was not favorable for the gas to particle conversion, whereas the frequent inversion conditions were likely favorable for the formation of SOC [40].

Daily variations of crustal elements are shown in Figure 6. Five crustal elements have a similar variation as the PM2.5 mass concentrations. Their concentrations varied significantly from day to day. The average concentration for Al, Si, Ca, Fe, and Ti was 0.51 ± 0.26, 1.01 ± 0.58, 0.74 ± 0.36, 1.10 ± 0.64, and 0.06 ± 0.03 $\mu g \cdot m^{-3}$, respectively. Increasing wind speed could be expected to increase the concentrations of crustal elements, but the concentration of the five elements had a weak correlation with the wind speed (R < 0.4) in the present study. However, when the wind speed exceeded 1.5 $m \cdot s^{-1}$ (11 and 18 November), the concentrations of crustal elements were higher than those on any other day. The FS mass concentration was estimated by summing the above five crustal elements plus oxygen for the normal oxides as Equation (1). The average mass concentrations of FS were 7.66 ± 3.62 $\mu g \cdot m^{-3}$, ranging from $1.99 – 15.1$ $\mu g \cdot m^{-3}$, and accounting for 9.42% of $PM_{2.5}$.

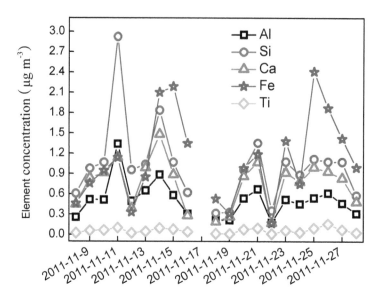

Figure 6. Time series of crustal elements concentration.

3.2. PM_{2.5} Mass Balance

The reconstructed $PM_{2.5}$ mass concentrations were close to the measured ones with strong correlation (Figure 7), indicating that the reconstruction of $PM_{2.5}$ could be reasonable. Nevertheless, a few biases were observed in the reconstructed $PM_{2.5}$ mass. Water absorption of the water-soluble components may lead to positive biases and overestimate the $PM_{2.5}$ mass, while the volatilization of NH_4NO_3 and volatile organic matter may result in negative biases. Moreover, the conversion used to estimate OM from OC also caused an uncertainty in calculating the $PM_{2.5}$ mass.

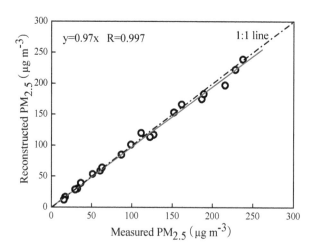

Figure 7. Scatter plots of measured and reconstructed PM_{2.5} mass concentrations.

Figure 8 presents the reconstructed chemical compositions in $PM_{2.5}$. On average, the fractions of major chemical compositions followed the order of OM > NH_4NO_3 > $(NH_4)_2SO_4$ > FS, CS, and EC. OM was the most abundant component in $PM_{2.5}$ (averaging 45.8 ± 31.7 $\mu g \cdot m^{-3}$), accounting for 39.4%

of PM₂.₅. The contribution of FS (averaging 7.66 ± 3.62 μg·m⁻³), CS (averaging 9.47 ± 6.24 μg·m⁻³), and EC (averaging 9.97 ± 6.28 μg·m⁻³) to PM₂.₅ was similar, each approximated to 9%. The percentage of SNA (30.2%) was much higher than the three species of FS, CS, and EC, but slightly lower than that of OM. The percentages of (NH₄)₂SO₄ and NH₄NO₃ (averaging 16.7 ± 13.2 and 19.0 ± 14.4 μg·m⁻³, respectively) were 14.9% and 15.4%, respectively. Compared with the results determined over the same period in earlier years [11], it is noted that the percentage of SNA in our study decreased by 3%–10% compared with that measured during 2003–2007, while the OM fraction rose about 5%.

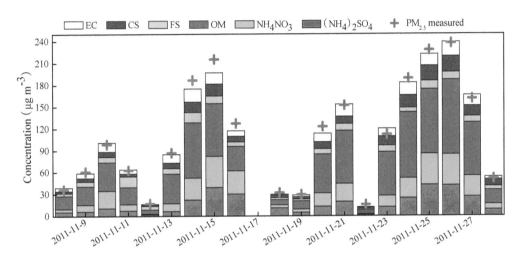

Figure 8. Daily variations of the reconstructed chemical composition in PM₂.₅.

3.3. Analysis of Aerosol Optical Properties

The time series of daily averaged optical properties including σ_{sc}, σ_{ap}, and SSA are shown in Figure 9. Daily σ_{sc} ranged from 31.1–667 Mm⁻¹, with a mean value of 270 ± 200 Mm⁻¹, while σ_{ap} was in the range of 8.24–158.0 Mm⁻¹, with a mean value of 74.3 ± 43.4 Mm⁻¹. The mean σ_{sc} value was considerable lower than that measured in urban Beijing in 2009 and during 2005–2006 [41,42], but higher than that obtained at a suburban site (Changping) and rural site (Shangdianzi) [15]. Compared with the results mentioned above, the mean σ_{ap} value was lower than that measured in 2009 as well, but higher than that during 2005–2006. The increased σ_{ap} in recent years is likely attributable to the rapid increase of vehicle pollution, since vehicular exhaust was one of the primary factors affecting aerosol absorption. The mean value of SSA was 0.76, which was comparable with the results determined in Beijing during 2005–2006 [42] and in 2009 [41].

MSE is an important parameter for estimating radiative forcing of aerosols and chemical extinction budgets for visibility impairment. Generally, there are two methods to estimate MSE, *i.e.*, measurement method and multilinear regression method [43]. MSE was defined as the ratio of measured σ_{sc} to aerosol mass concentration according to the measurement method. One alternative method can also be used by regression of the measured σ_{sc} against aerosol mass concentration. Since the RH in the nephelometer was maintained below 60% and the PM₂.₅ mass concentrations were measured at RH of 40%, those days with RH below 50% were selected for regressing to minimize the impact of particle hygroscopic growth on σ_{sc}. According to the measurement method, daily MSE varied from 1.70–3.02 m²·g⁻¹, with a mean value of 2.32 ± 0.44 m²·g⁻¹. It is noted that a strong correlation between the measured σ_{sc} and PM$_{2.5}$ mass concentration was observed with a high correlation coefficient of 0.988 (Figure 9). The slope was 2.67 from a linear regression, which was slightly higher than the value obtained by the measurement method, also found by Titos *et al.* [43]. Compared to that measured in urban Beijing by Zhao *et al.*, and Jing *et al.* [41], the relative lower MSE in our study was likely related to the heavily polluted events.

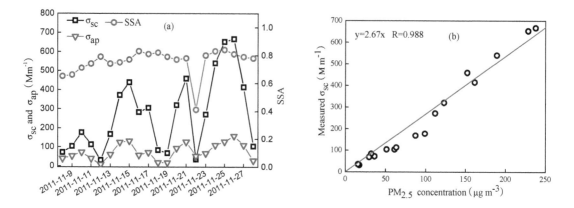

Figure 9. The time series of daily averaged aerosol optical properties (**a**) and relationship between the measured σ_{sc} and PM$_{2.5}$ concentration (**b**).

3.4. Chemical Apportionment of the Aerosol Optical Parameters

In order to appoint the contribution to the visibility impairment, b$_{ext}$ was reconstructed based on the chemical compositions of aerosol. In the present study, the extinction effect by fine particles was studied, while the contributions of gases were excluded because they only accounted for a small fraction (about 2%–4%) of b$_{ext}$ [44]. The impact of sea salt on b$_{ext}$ was ignored since Beijing is about 150 km away from the East China's coastal oceans. Moreover, the contribution of coarse mass to b$_{ext}$ was not included because of lack of the concentrations of coarse matter. Then, the revised IMPROVE formula of Equation (2) was modified as follows:

$$
\begin{aligned}
b_{ext} = {} & 2.2 \times f_s(RH) \times [Small(NH_4)_2 SO_4] + 4.8 \times f_L(RH) \times [L \arg e(NH_4)_2 SO_4] \\
& + 2.4 \times f_s(RH) \times [SmallNH_4NO_3] + 5.1 \times f_L(RH) \times [L \arg eNH_4NO_3] \\
& + 2.8 \times [SmallOM] + 6.1 \times [L \arg eOM] \\
& + 10 \times [EC] + 1 \times [Fine\ Soil]
\end{aligned}
\tag{11}
$$

The measured and reconstructed b$_{ext}$ are illustrated in Figure 10. It is found that the measured b$_{ext}$ were considerably lower than the reconstructed value, especially during the heavily pollution levels. The deviation varied from 18.1%–140%, with an average value of about 70%. Jung *et al.* [45] also found that the b$_{ext}$ was overestimated by 36.7% based on the revised IMPROVE algorithm. However, a few other studies using the IMPROVE formula found that there existed a good correlation between the measured and reconstructed b$_{ext}$, and the slopes were close to 1.0 in Shanghai [19] and Guangzhou [4]. Compared to the results conducted in Shanghai, a lower MAE (7.7) was used to calculate b$_{ap}$. If the same value of 7.7 was used in the present study, the biases would decrease by 8%. Additionally, the MSE used to calculate b$_{ext}$ in Guangzhou was much higher than the value in the present study. Thus, it can be deduced that a lower MSE than the value adopted according to the revised IMPROVE algorithm in the present study should be used to reconstruct b_{ext}, which may result in a reconstructed b_{ext} approximately equal to the measured b_{ext}. Although more locally-derived MSE and MAE were necessary for effectively reconstructing the b_{ext}, we did not obtain these values in the present study due to lack of the amount of *in situ* and sampling data.

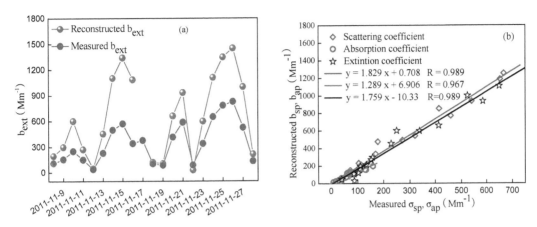

Figure 10. The temporal variations of measured and reconstructed b_{ext} (**a**) and the correlation between measured and reconstructed optical parameters (**b**).

Although the reconstructed b_{ext} was higher than that measured one, both were correlated well (R = 0.989), so did b_{sp} and b_{ap} (Figure 11). Therefore, the relative contribution of each chemical composition to b_{ext} can also be analyzed by the modified IMPROVE algorithm. As shown in Figure 11, OM, $(NH_4)_2SO_4$, NH_4NO_3 and EC were the dominant contributors, accounting for 97.7% of b_{ext} together, while the contribution of FS was small, accounting for only 2.3% of b_{ext}. On average, OM was the largest contributor to the b_{ext}, accounting for 44.7% of b_{ext}, while SNA accounted for 34.4% of b_{ext}. $(NH_4)_2SO_4$ and NH_4NO_3 contributed 16.4% and 18.0% of b_{ext}, respectively. Our results were different from those determined in Beijing in previous studies conducted in summer [16,18,46], which showed that $(NH_4)_2SO_4$ and NH_4NO_3 were the largest contributor to the b_{ext}.

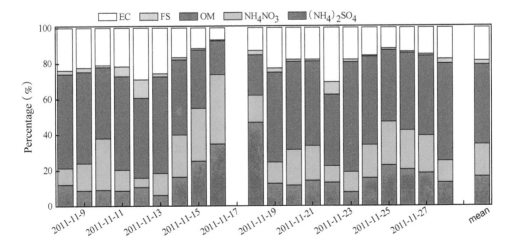

Figure 11. Relative contributions of each chemical composition in PM2.5 to b_{ext}.

3.5. Typical Pollution Episodes

As shown in Figure 1, four obvious pollution episodes were observed during the campaign, with the visibility deteriorating to less than 10 km. They were observed on 9–11 November, 14–17 November, 20 and 21 November, and 23–27 November. Obviously, in the first pollution period, pollutants accumulated gradually from 8–10 November, with daily $PM_{2.5}$ mass concentration increasing from 36.3–98.8 $\mu g \cdot m^{-3}$, and then decreasing dramatically to 17.1 $\mu g \cdot m^{-3}$. All the chemical components increased with $PM_{2.5}$ mass, especially for NO_3^-, which increased by six times compared with the value on 8 November. Furthermore, it can be found that during the pollution accumulation period, the wind speed was less than 0.5 $m \cdot s^{-1}$. On 11 November, strong wind was favorable for dispersion of the pollutants and accumulation of the crustal material. Moreover, 12 November was weekend, and the reduction of vehicles may also contribute to the lower concentration of $PM_{2.5}$. From an extinction perspective, in the first pollution stage, σ_{sc} increased from 71.6–177.2 Mm^{-1}, while σ_{ap} varied little, implying that the visibility degradation was mainly caused by the rapid increase of SNA.

In the second pollution period from 12–19 November, daily $PM_{2.5}$ mass concentration increased from 17.1–215.5 $\mu g \cdot m^{-3}$, with the maximum value occurring on 15 November, and then plunging to 32.0 $\mu g \cdot m^{-3}$ within three days. As illustrated in Figure 1, it can be found that the wind speeds were very low from 12–15 November, and in favor of the accumulation of the pollutants. On the other hand, the RH increased from 30% to 82%, which was favorable for the formation of SNA. Residential heating starting in the middle of November might be the primary reason for the heavier pollution in the second stage. As shown in Figure 2, the concentration of Cl^- on 13 November had a dramatic increase compared with the previous days, approximately to 10 times higher than that on 12 November. Chloride may be essentially contributed by coal combustion in Beijing during the heating season. Thus, the high $PM_{2.5}$ mass concentration was mainly associated with coal combustion. In fact, the concentrations of OC and EC during 13–27 November were much higher than those before 12 November. The sharp increase of OC and EC also verified the influence of coal combustion on the increase of $PM_{2.5}$ mass concentration. Unlike the first pollution stage, σ_{ap} on the most heavy pollution day (15 November) was about 15 times higher than that on the clean days. Meanwhile, σ_{sc} rose from 31.0–439.2 Mm^{-1}, with a similar growth rate as σ_{ap}. As shown in Figure 11, during the second pollution episode, the contribution of OM and EC to b_{ext} decreased from 54.2% and 25.6% to 32.7% and 11.5% as the pollutants accumulated, respectively, while the contribution of SNA to b_{ext} increased from 18.3%–54.8% on the accumulation period. On 16 November, although the mass concentration of $PM_{2.5}$ reduced by 40%, the visibility was still less than 3 km, which was ascribed to the largest contribution of SNA to b_{ext}. As presented in Figure 11, the contribution of SNA reached up to 73.7% whereas that of OM and EC decreased to 26.7% to b_{ext}. The $PM_{2.5}$ mass concentration had a significant decrease on 18 November due to rain and strong wind (Figure 1).

Based on the analysis of a typical pollution episode, it can be concluded that the secondary formation of aerosol was more important than the carbonaceous pollution for the haze formation in Beijing. In the other two pollution periods, a similar trend of the chemical composition to that during the second pollution stage was observed. In general, the pollution accumulation was in accordance with the increase of the SNA, OC and EC concentrations under stable weather conditions until arrival of strong wind.

4. Conclusions

During the heating period from 8–28 November 2011, aerosol optical properties as well as chemical compositions were investigated simultaneously in Beijing. Daily $PM_{2.5}$ mass concentration varied from 15.6–237.5 $\mu g \cdot m^{-3}$ and presented a mean value of 111.2 ± 73.4 $\mu g \cdot m^{-3}$. Among the chemical components in $PM_{2.5}$, NO_3^- was the most abundant species in WSIIs with an average concentration of 14.7 ± 11.2 $\mu g \cdot m^{-3}$, followed by SO_4^{2-}, NH_4^+, and Cl^-, accounting for 28.9%, 25.6%, 17.7%, and 14.6% of WSIIs, respectively. The rest of K^+, Na^+, Ca^{2+}, F^-, and Mg^{2+} have a minor contribution to the WSIIs, accounting for 13.2% of WSIIs together.

The mean σ_{sc}, σ_{ap} and SSA values at 550 nm were 270 ± 200 Mm^{-1}, 74.3 ± 43.4 Mm^{-1} and 0.76 during the entire observation period, respectively. Both of the σ_{sc} and σ_{ap} increased significantly during the pollution accumulation episode. The b_{ext} were estimated by the revised IMPROVE formula based on the chemical compositions of $PM_{2.5}$. Compared with the measured σ_{sc} and σ_{ap}, the reconstructed b_{ext} was overestimated, but had a strong correlation with a high correlation coefficient of 0.989. OM was the largest contributor, accounting for 44.7% of b_{ext}, followed by NH_4NO_3, $(NH_4)_2SO_4$, with minor contribution from soil dust (2.3%).

Pollution episodes in Beijing were strongly influenced by both emissions and meteorological conditions. Pollutant was accumulated in calm or weak winds while diffused under strong wind conditions. Additionally, the coal combustion for residential heating was another major reason for the heavy pollution during the sampling period. Four typical pollution episodes during the study period were observed, it was found that NH_4NO_3 and $(NH_4)_2SO_4$ were the largest contributor to the b_{ext} rather than carbonaceous components during the pollution accumulation episodes, implying that the secondary inorganic pollutants were more important than the carbonaceous pollution for heavy pollution formation. Therefore, the reduction of their precursors such as SO_2, NO_x and NH_3 could effectively improve the visibility in Beijing.

Acknowledgments

This study was supported by the National Natural Science Foundation of China projects (41075093, 41275121 and 41375123), the "Strategic Priority Research Program" of the Chinese Academy of Sciences (KJZD-EW-TZ-G06-04), the Ministry of Environmental Protection of China (201209007), State Environmental Protection Key Laboratory of Sources and Control of Air Pollution Complex (SCAPC201310), Jiangsu Key Laboratory of Atmospheric Environment Monitoring and Pollution Control of Nanjing University of Information Science and Technology, and Jiangsu Province Innovation Platform for Superiority Subject of Environmental Science and Engineering (KHK1201). The authors thank Lian-fang Wei, Jin-lu Dong, and Rong Zhang for their contributions to the field and laboratory work.

Author Contributions

The work was completed with collaboration between all the authors. The corresponding author designed the research theme, organized the $PM_{2.5}$ sampling with Xinghua Li, checked the experimental results, and designed the manuscript with Huanbo Wang. Huanbo Wang analyzed the data, interpreted the results and wrote the manuscript. Xinghua Li was in charge of $PM_{2.5}$ sampling and collected all relevant data, and Chengcai Li was in charge of observation of optical parameters. Junji Cao was in charge of inorganic elements analysis. Yongliang Ma and Kebin He provided analyses of water-soluble ions and OC/EC. Guangming Shi was involved in relevant data interpretation and discussion.

Conflicts of Interest

The authors declare no conflict of interest.

Reference

1. Wang, Q.Y.; Cao, J.J.; Tao, J.; Li, N.; Su, X.O.; Chen, L.W.A.; Wang, P.; Shen, Z.X.; Liu, S.X.; Dai, W.T. Long-term trends in visibility and at Chengdu, China. *PLoS One* **2013**, doi:10.1371/journal.pone.0068894

2. Zhang, X.Y.; Wang, Y.Q.; Niu, T.; Zhang, X.C.; Gong, S.L.; Zhang, Y.M.; Sun, J.Y. Atmospheric aerosol compositions in China: Spatial/temporal variability, chemical signature, regional haze distribution and comparisons with global aerosols. *Atmos. Chem. Phys.* **2012**, *12*, 779–799.

3. Shen, G.F.; Xue, M.; Yuan, S.Y.; Zhang, J.; Zhao, Q.Y.; Li, B.; Wu, H.S.; Ding, A.J. Chemical compositions and reconstructed light extinction coefficients of particulate matter in a mega-city. In the western Yangtze River Delta, China. *Atmos. Environ.* **2014**, *83*, 14–20.

4. Tao, J.; Zhang, L.M.; Ho, K.F.; Zhang, R.J.; Lin, Z.J.; Zhang, Z.S.; Lin, M.; Cao, J.J.; Liu, S.X.; Wang, G.H. Impact of $PM_{2.5}$ chemical compositions on aerosol light scattering in Guangzhou—The largest megacity in south China. *Atmos. Res.* **2014**, *135*, 48–58.

5. Chen, J.; Qiu, S.S.; Shang, J.; Wilfrid, O.M.F.; Liu, X.G.; Tian, H.Z.; Boman, J. Impact of relative humidity and water soluble constituents of $PM_{2.5}$ on visibility impairment in Beijing, China. *Aerosol Air Qual. Res.* **2014**, *14*, 260–268.

6. Ramanathan, V.; Feng, Y. Air pollution, greenhouse gases and climate change: Global and regional perspectives. *Atmos. Environ.* **2009**, *43*, 37–50.

7. Ma, X.; Yu, F.; Luo, G. Aerosol direct radiative forcing based on GEOS-Chem-APM and uncertainties. *Atmos. Chem. Phys.* **2012**, *12*, 5563–5581.

8. Yang, F.; Huang, L.; Duan, F.; Zhang, W.; He, K.; Ma, Y.; Brook, J.R.; Tan, J.; Zhao, Q.; Cheng, Y. Carbonaceous species in $PM_{2.5}$ at a pair of rural/urban sites in Beijing, 2005–2008. *Atmos. Chem. Phys.* **2011**, *11*, 7893–7903.

9. Yang, F.M.; He, K.B.; Ma, Y.L.; Zhang, Q.; Cadle, S.H.; Chan, T.; Mulawa, P.A. Characterization of carbonaceous species of ambient $PM_{2.5}$ in Beijing, China. *J. Air Waste Manag.* **2005**, *55*, 984–992.

10. Zhao, P.S.; Dong, F.; He, D.; Zhao, X.J.; Zhang, X.L.; Zhang, W.Z.; Yao, Q.; Liu, H.Y. Characteristics of concentrations and chemical compositions for $PM_{2.5}$ in the region of Beijing, Tianjin, and Hebei, China. *Atmos. Chem. Phys.* **2013**, *13*, 4631–4644.

11. Yang, F.; Tan, J.; Zhao, Q.; Du, Z.; He, K.; Ma, Y.; Duan, F.; Chen, G.; Zhao, Q. Characteristics of $PM_{2.5}$ speciation in representative megacities and across China. *Atmos. Chem. Phys.* **2011**, *11*, 5207–5219.

12. Zhang, R.; Jing, J.; Tao, J.; Hsu, S.C.; Wang, G.; Cao, J.; Lee, C.S.L.; Zhu, L.; Chen, Z.; Zhao, Y.; *et al.* Chemical characterization and source apportionment of PM$_{2.5}$ in Beijing: Seasonal perspective. *Atmos. Chem. Phys.* **2013**, *13*, 7053–7074.

13. Xu, J.W.; Tao, J.; Zhang, R.J.; Cheng, T.T.; Leng, C.P.; Chen, J.M.; Huang, G.H.; Li, X.; Zhu, Z.Q. Measurements of surface aerosol optical properties in winter of Shanghai. *Atmos. Res.* **2012**, *109*, 25–35.

14. Yan, P.; Tang, J.; Huang, J.; Mao, J.T.; Zhou, X.J.; Liu, Q.; Wang, Z.F.; Zhou, H.G. The measurement of aerosol optical properties at a rural site in northern China. *Atmos. Chem. Phys.* **2008**, *8*, 2229–2242.

15. Zhao, X.J.; Zhang, X.L.; Pu, W.W.; Meng, W.; Xu, X.F. Scattering properties of the atmospheric aerosol in Beijing, China. *Atmos. Res.* **2011**, *101*, 799–808.

16. Jung, J.; Lee, H.; Kim, Y.J.; Liu, X.G.; Zhang, Y.H.; Hu, M.; Sugimoto, N. Optical properties of atmospheric aerosols obtained by *in situ* and remote measurements during 2006 Campaign of Air Quality Research in Beijing (CAREBeijing-2006). *J. Geophys. Res.: Atmos.* **2009**, doi:10.1029/2008JD010337.

17. Garland, R.M.; Yang, H.; Schmid, O.; Rose, D.; Nowak, A.; Achtert, P.; Wiedensohler, A.; Takegawa, N.; Kita, K.; Miyazaki, Y.; *et al.* Aerosol optical properties in a rural environment near the mega-city Guangzhou, China: Implications for regional air pollution, radiative forcing and remote sensing. *Atmos. Chem. Phys.* **2008**, *8*, 5161–5186.

18. Tian, P.; Wang, G.F.; Zhang, R.J. Impacts of aerosol chemical compositions on optical properties in urban Beijing, China. *Particuology* **2014**, doi:10.1016/j.partic.2014.1003.1014.

19. Huang, G.H.; Cheng, T.T.; Zhang, R.J.; Tao, J.; Leng, C.P.; Zhang, Y.W.; Zha, S.P.; Zhang, D.Q.; Li, X.; Xu, C.Y. Optical properties and chemical composition of PM$_{2.5}$ in Shanghai in the spring of 2012. *Particuology* **2014**, *13*, 52–59.

20. Pitchford, M.; Malm, W.; Schichtel, B.; Kumar, N.; Lowenthal, D.; Hand, J. Revised algorithm for estimating light extinction from improve particle speciation data. *J. Air Waste Manag.* **2007**, *57*, 1326–1336.

21. Chow, J.C.; Watson, J.G.; Chen, L.W.A.; Chang, M.C.O.; Robinson, N.F.; Trimble, D.; Kohl, S. The IMPROVE_A temperature protocol for thermal/optical carbon analysis: Maintaining consistency with a long-term database. *J. Air Waste Manag.* **2007**, *57*, 1014–1023.

22. Xu, H.M.; Cao, J.J.; Ho, K.F.; Ding, H.; Han, Y.M.; Wang, G.H.; Chow, J.C.; Watson, J.G.; Khol, S.D.; Qiang, J.; *et al.* Lead concentrations in fine particulate matter after the phasing out of leaded gasoline in Xi'an, China. *Atmos. Environ.* **2012**, *46*, 217–224.

23. Xing, L.; Fu, T.M.; Cao, J.J.; Lee, S.C.; Wang, G.H.; Ho, K.F.; Cheng, M.C.; You, C.F.; Wang, T.J. Seasonal and spatial variability of the OM/OC mass ratios and high regional correlation between oxalic acid and zinc in Chinese urban organic aerosols. *Atmos. Chem. Phys.* **2013**, *13*, 4307–4318.

24. Malm, W.C.; Sisler, J.F.; Huffman, D.; Eldred, R.A.; Cahill, T.A. Spatial and seasonal trends in particle concentration and optical extinction in the United States. *J. Geophys. Res.: Atmos.* **1994**, *99*, 1347–1370.

25. Barnard, J.C.; Kassianov, E.I.; Ackerman, T.P.; Johnson, K.; Zuberi, B.; Molina, L.T.; Molina, M.J. Estimation of a "radiatively correct" black carbon specific absorption during the Mexico City Metropolitan Area (MCMA) 2003 field campaign. *Atmos. Chem. Phys.* **2007**, *7*, 1645–1655.

26. Zhou, J.M.; Zhang, R.J.; Cao, J.J.; Chow, J.C.; Watson, J.G. Carbonaceous and ionic components of atmospheric fine particles in Beijing and their impact on atmospheric visibility. *Aerosol Air Qual. Res.* **2012**, *12*, 492–502.

27. Sun, K.; Qu, Y.; Wu, Q.; Han, T.T.; Gu, J.W.; Zhao, J.J.; Sun, Y.L.; Jiang, Q.; Gao, Z.Q.; Hu, M.; *et al.* Chemical characteristics of size-resolved aerosols in winter in Beijing. *J. Environ. Sci.* **2014**, *26*, 1641–1650.

28. Arimoto, R.; Duce, R.A.; Savoie, D.L.; Prospero, J.M.; Talbot, R.; Cullen, J.D.; Tomza, U.; Lewis, N.F.; Jay, B.J. Relationships among aerosol constituents from Asia and the North Pacific during Pem-West A. *J. Geophys. Res.: Atmos.* **1996**, *101*, 2011–2023.

29. Duan, F.K.; He, K.B.; Ma, Y.L.; Yang, F.M.; Yu, X.C.; Cadle, S.H.; Chan, T.; Mulawa, P.A. Concentration and chemical characteristics of PM$_{2.5}$ in Beijing, China: 2001–2002. *Sci. Total Environ.* **2006**, *355*, 264–275.

30. Wang, S.X.; Xing, J.; Chatani, S.; Hao, J.M.; Klimont, Z.; Cofala, J.; Amann, M. Verification of anthropogenic emissions of China by satellite and ground observations. *Atmos. Environ.* **2011**, *45*, 6347–6358.

31. Wang, Y.; Zhuang, G.S.; Tang, A.H.; Yuan, H.; Sun, Y.L.; Chen, S.A.; Zheng, A.H. The ion chemistry and the source of PM$_{2.5}$ aerosol in Beijing. *Atmos. Environ.* **2005**, *39*, 3771–3784.

32. Wang, R.S.; Xiao, H.; Bai, T.; Wang, Y.J.; Qian, L.Y. The amount of vehicles by "China vehicle emission control annual report in 2013". *Environ. Sustain. Dev.* **2014**, *2014*, 88–90.

33. Cheng, Y.; Engling, G.; He, K.B.; Duan, F.K.; Ma, Y.L.; Du, Z.Y.; Liu, J.M.; Zheng, M.; Weber, R.J. Biomass burning contribution to Beijing aerosol. *Atmos. Chem. Phys.* **2013**, *13*, 7765–7781.

34. Yang, F.; He, K.; Ye, B.; Chen, X.; Cha, L.; Cadle, S.H.; Chan, T.; Mulawa, P.A. One-year record of organic and elemental carbon in fine particles in downtown Beijing and Shanghai. *Atmos. Chem. Phys.* **2005**, *5*, 1449–1457.

35. Cao, J.J.; Lee, S.C.; Chow, J.C.; Watson, J.G.; Ho, K.F.; Zhang, R.J.; Jin, Z.D.; Shen, Z.X.; Chen, G.C.; Kang, Y.M.; *et al.* Spatial and seasonal distributions of carbonaceous aerosols over China. *J. Geophys. Res.: Atmos.* **2007**, doi:10.1029/2006JD008205.

36. Turpin, B.J.; Huntzicker, J.J. Identification of secondary organic aerosol episodes and quantitation of primary and secondary organic aerosol concentrations during SCAQS. *Atmos. Environ.* **1995**, *29*, 3527–3544.

37. Watson, J.G.; Chow, J.C.; Houck, J.E. PM$_{2.5}$ chemical source profiles for vehicle exhaust, vegetative burning, geological material, and coal burning in northwestern Colorado during 1995. *Chemosphere* **2001**, *43*, 1141–1151.

38. Chow, J.C.; Watson, J.G.; Lu, Z.Q.; Lowenthal, D.H.; Frazier, C.A.; Solomon, P.A.; Thuillier, R.H.; Magliano, K. Descriptive analysis of PM$_{2.5}$ and PM$_{10}$ at regionally representative locations during SJVAQS/AUSPEX. *Atmos. Environ.* **1996**, *30*, 2079–2112.

39. Castro, L.M.; Pio, C.A.; Harrison, R.M.; Smith, D.J.T. Carbonaceous aerosol in urban and rural European atmospheres: Estimation of secondary organic carbon concentrations. *Atmos. Environ.* **1999**, *33*, 2771–2781.

40. Sheehan, P.E.; Bowman, F.M. Estimated effects of temperature on secondary organic aerosol concentrations. *Environ. Sci. Technol.* **2001**, *35*, 2129–2135.

41. Jing, J.S.; Wu, Y.F.; Tao, J.; Che, H.Z. Observation and analysis of near-surface atmospheric aerosol optical properties in urban Beijing. *Particuology* **2014**, doi:10.1016/j.partic.2014.1003.1013.

42. He, X.; Li, C.C.; Lau, A.K.H.; Deng, Z.Z.; Mao, J.T.; Wang, M.H.; Liu, X.Y. An intensive study of aerosol optical properties in Beijing urban area. *Atmos. Chem. Phys.* **2009**, *9*, 8903–8915.

43. Titos, G.; Foyo-Moreno, I.; Lyamani, H.; Querol, X.; Alastuey, A.; Alados-Arboledas, L. Optical properties and chemical composition of aerosol particles at an urban location: An estimation of the aerosol mass scattering and absorption efficiencies. *J. Geophys. Res.: Atmos.* **2012**, doi:10.1029/2011JD016671.

44. Cao, J.J.; Wang, Q.Y.; Chow, J.C.; Watson, J.G.; Tie, X.X.; Shen, Z.X.; Wang, P.; An, Z.S. Impacts of aerosol compositions on visibility impairment in Xi'an, China. *Atmos. Environ.* **2012**, *59*, 559–566.

45. Jung, J.; Lee, H.; Kim, Y.J.; Liu, X.G.; Zhang, Y.H.; Gu, J.W.; Fan, S.J. Aerosol chemistry and the effect of aerosol water content on visibility impairment and radiative forcing in Guangzhou during the 2006 Pearl River Delta campaign. *J. Environ. Manag.* **2009**, *90*, 3231–3244.

46. Li, X.H.; He, K.B.; Li, C.C.; Yang, F.M.; Zhao, Q.; Ma, Y.L.; Cheng, Y.; Ouyang, W.J.; Chen, G.C. PM$_{2.5}$ mass, chemical composition, and light extinction before and during the 2008 Beijing Olympics. *J. Geophys. Res.: Atmos.* **2013**, *118*, 12158–12167.

8

Role of Surface Energy Exchange for Simulating Wind Turbine Inflow: A Case Study in the Southern Great Plains, USA

Sonia Wharton [1,*], Matthew Simpson [1], Jessica L. Osuna [1], Jennifer F. Newman [2] and Sebastien C. Biraud [3]

[1] Atmospheric, Earth and Energy Division, Lawrence Livermore National Laboratory, 7000 East Avenue, L-103, Livermore, CA 94550, USA; E-Mails: simpson35@llnl.gov (M.S.); osuna2@llnl.gov (J.O.)

[2] School of Meteorology, University of Oklahoma, 120 David L. Boren Blvd., Norman, OK 73072, USA; E-Mail: jennifer.newman@ou.edu

[3] Earth Sciences Division, Lawrence Berkeley National Laboratory, 1 Cyclotron Rd., Berkeley, CA 94720, USA; E-Mail: scbiraud@lbl.gov

* Author to whom correspondence should be addressed; E-Mail: wharton4@llnl.gov

Academic Editor: Stephan De Wekker

Abstract: The Weather Research and Forecasting (WRF) model is used to investigate choice of land surface model (LSM) on the near surface wind profile, including heights reached by multi-megawatt (MW) wind turbines. Simulations of wind profiles and surface energy fluxes were made using five LSMs of varying degrees of sophistication in dealing with soil–plant–atmosphere feedbacks for the Department of Energy (DOE) Southern Great Plains (SGP) Atmospheric Radiation Measurement Program (ARM) Central Facility in Oklahoma, USA. Surface flux and wind profile measurements were available for validation. WRF was run for three, two-week periods covering varying canopy and meteorological conditions. The LSMs predicted a wide range of energy flux and wind shear magnitudes even during the cool autumn period when we expected less variability. Simulations of energy fluxes varied in accuracy by model sophistication, whereby LSMs with very simple or no soil–plant–atmosphere feedbacks were the least accurate; however, the most complex models did not consistently produce more accurate results. Errors in wind shear were also sensitive to LSM choice and were partially related to energy flux accuracy. The variability of LSM performance was relatively high suggesting that LSM representation of energy fluxes in WRF remains a large source of model uncertainty for simulating wind turbine inflow conditions.

Keywords: WRF; land surface model; Southern Great Plains; ARM; surface energy fluxes; lidar; wind energy; wind shear

1. Introduction

Atmospheric models are not perfect predictors of incoming wind conditions or "inflow" at heights spanned by industrial-scale wind turbines (~40 to 200 m above ground level, a.g.l.). One way to optimize model accuracy is to identify areas in numerical models that may lead to a wind forecasting improvement. This is a warranted endeavor, as a wind forecasting improvement of as little as 10%–20% could result in hundreds of millions of dollars ($140M–$260M) in annual operating cost savings for the U.S. wind industry [1]. One candidate for optimization is the user's choice of land surface model (LSM) employed in numerical weather prediction models. The LSM controls the exchange of energy between the surface and the atmosphere and may have a large effect on inflow in the lower boundary layer. We hypothesize that wind speeds simulated at heights equivalent to a turbine rotor disk are sensitive to LSM choice due to variations in the sophistication of each model's characterization or parameterization of the soil–vegetation–atmosphere continuum. In this work, this theory is tested for northern Oklahoma using the Weather Research and Forecasting (WRF) model [2].

Oklahoma is currently ranked 6th in the nation for total megawatts installed (over 3300 MW) and is one of the leading states for percentage of electricity provided by wind (nearly 15%) [3]. As such, the Department of Energy (DOE) Southern Great Plains (SGP) Atmospheric Radiation Measurement Program (ARM) Central Facility in northern Oklahoma in recent years has become an ideal location for conducting wind energy research. Moreover, in 2013, a multi-MW wind farm began operation just to the west of the ARM Central Facility. At present, this farm has 140 General Electric (GE) 1.68 MW turbines, each with a hub-height of 80 m, and rotor diameter of 82.5 m. Each turbine blade tip has a minimum height of 39 m and a maximum height of 121 m and is located well within the lower PBL. The farm rotor diameters are typical of the average size (89 m) installed at present in the U.S. [4].

WRF is heavily utilized in the atmospheric community for weather prediction and more recently for wind forecasting, including its use in predicting hub-height wind speed and direction in the wind power industry. WRF has many physics options that are user defined, including options for choosing the planetary boundary layer (PBL) scheme, surface layer scheme, radiation scheme, microphysics and convective schemes, and LSM. Given the large number LSMs currently available in WRF, it remains a challenge for modelers to select and configure the appropriate land surface scheme to fit their needs [5]. These models range from those with a simplistic treatment of surface processes (e.g., no plant canopy) to sophisticated subsurface-vegetation-atmosphere transfer models. The LSM in WRF calculates heat and moisture fluxes over land to provide a lower boundary condition for vertical transport (*i.e.*, atmospheric mixing) in the PBL scheme.

Energy exchange at the surface can strongly drive lower PBL flow features in the daytime, as thermally-generated, buoyant sensible heat fluxes, in combination with mesoscale forcing and drag across the surface roughness elements, largely determine the magnitude of wind shear (*i.e.*, change in wind speed with height) and turbulence at heights encountered by a wind turbine rotor-disk. Previous studies have examined the role of surface and subsurface properties on near-surface meteorology and PBL development (e.g., [6–13]), the role of the PBL scheme in WRF for wind forecasting (e.g., [14,15]), and the impact of crop type and surface drag (via the aerodynamic roughness length, z_o) on hub-height wind speed and shear [16]. However, few studies have specifically looked at the role of surface energy exchange and LSM choice on rotor-disk wind shear. This is a potential area for wind forecasting improvement as numerous studies have shown that wind shear characteristics are in part dependent on atmospheric stability and that shear has significant impacts on turbine power generation (e.g., [17–23]).

Surface energy exchange is described by the terrestrial radiation budget, defined in Equation (1), where R_n is net radiation flux ($W \cdot m^{-2}$) (*i.e.*, the difference between incoming and outgoing radiation), H is sensible heat flux ($W \cdot m^{-2}$), LE is latent energy flux ($W \cdot m^{-2}$), G is ground heat flux ($W \cdot m^{-2}$) and S is biomass storage heat flux ($W \cdot m^{-2}$).

$$R_n = H - LE - G - S \tag{1}$$

While some of the available energy is absorbed by the ground and biomass, this is on average less than 15% of the net radiation flux for most plant canopies and the majority of available energy is transferred back into the atmosphere as sensible and latent heat [8]. Latent heat is the quantity of heat absorbed or released by water undergoing a change of state. Over a plant canopy, LE is the heat released by water as it changes from a liquid to gaseous state through evapotranspiration (ET). ET includes water evaporated from the canopy and ground surfaces plus water lost through gas exchange via the plant vascular system (transpiration).

The magnitudes of H, LE, G and S are dependent on many variables as well as the nature of the feedbacks between them. These variables include the amount of incoming radiation, soil moisture availability, groundwater availability and plant access, soil properties, canopy properties (e.g., species, albedo, biomass density, leaf area index (LAI), and rooting depth), air and soil temperature, relative humidity, and entrainment of dry air into the boundary layer from the free atmosphere. The ratio of sensible heat to latent energy transfer is called the Bowen ratio ($\beta = H/LE$) and is usually expressed as a midday average. A high β indicates that a larger portion of available energy is partitioned into sensible heat than latent heat.

Here we largely focus on the partitioning of incoming radiation into sensible and latent energy. Wind shear at heights encountered by wind turbines is hypothesized to be strongly correlated to β during the summer months when land-atmosphere interactions are the strongest. Large H fluxes produce strongly buoyant heat fluxes which increase thermally-driven mixing in the lower PBL and decrease wind shear. On the other hand, smaller H fluxes will lead to weaker thermally-driven mixing and an environment with possibly greater wind shear across the wind turbine rotor-disk. H fluxes can be attenuated for a number of reasons. For example, if they are relatively small during the warm summer months this is usually because a large amount of available energy is going into evaporation and transpiration of water from the crop canopy. Despite the significant influence of H and LE on the daytime PBL, uncertainty remains in the parameterization of surface heat and moisture fluxes in numerical weather prediction models [24].

Experimental studies such as CASES-97 and IHOP-2002 have shown that latent and sensible heat fluxes in the SGP region vary significantly in time and space due to land use, climatology, and soil type differences across the landscape [25–27]. However, a typical wind forecaster, such as a wind plant operator, usually runs WRF in a "standard mode" which does not capture fine-scale land use heterogeneity nor fine-scale temporal vegetation changes. LSMs in standard WRF use either a monthly or seasonally-adjusted or annually-fixed land use value from look-up tables to assign surface characteristics, but crop cover can change frequently and suddenly in this region; for example, rapid canopy changes would be caused by crop "greening" or senescence, flood or hail crop losses, or harvest or field tillage events. This coarse approach for incorporating land use and vegetation property changes in LSMs is likely a shortcoming in areas such as the SGP with high spatial and temporal surface variability. To improve modeling in heterogeneous areas, some research groups have incorporated high-resolution remote sensing surface data into their land surface model [28,29]. This "mosaic approach" of using satellite products for data assimilation has been incorporated into certain, research-geared applications of LSMs such as Noah, RUC and the Community Land Model (CLM).

While satellite data assimilation of near real-time surface properties would likely result in more accurate surface fluxes, and potentially more accurate simulations of turbine-relevant wind inflow conditions, this was not employed here. In brief, we wanted to evaluate the accuracy of each LSM in WRF by running WRF in a "standard" mode as a wind plant operator or forecaster would do. This is done for a number of reasons. First, wind plant operators are usually limited by their computational resources which restrict them to run WRF in mesoscale mode and not at finer resolution. Second, plant operators need to run forecasts quickly as the energy market relies on hourly and day ahead forecasts. The recent operation of a neighboring wind farm provided additional motivation for studying the importance of LSMs in a standard version of WRF.

Here we aim to answer the following questions for a wide range of LSMs and surface and meteorological conditions:

(1) Is there a correlation between surface energy exchange and wind shear in the observations at this site? Is this correlation seasonal?
(2) Is modeled energy flux and wind shear accuracy related to complexity in the LSM's parameterization of soil–vegetation–atmosphere feedbacks?

2. Methods

2.1. Site Description and Instrumentation

All measurements were taken at the DOE SGP ARM Central Facility near Lamont, Oklahoma (36.605°N, 97.488°W, 317 m a.s.l.) (Figure 1). SGP ARM is set in a rural landscape dominated by pastureland and annually grown crops. The landscape is relatively flat with a local terrain slope less than three degrees. The species of annual crops planted in fields adjacent to the Central Facility vary across years and seasons depending on climatic conditions and market prices. Crops can include winter wheat and canola, typically grown from fall to early summer, and corn, sorghum, cowpeas, barley and soybeans grown during spring and summer. The region has fine scale spatial heterogeneity due to small land holdings and changing crop rotations. Regional heterogeneity is enhanced by climatic gradients resulting in a wide range of precipitation across the larger SGP ARM domain from the northwest (380 mm) to the southeast (1270 mm) [30]. The crops are not irrigated and crop losses, delayed plantings, and early or late harvests can result from anomalous climate conditions. Area soils are classified as well-drained silt loam, silty clay loam or clay loam and have permanent wilting points of ~0.13 $m^3 \cdot m^{-3}$. Permanent wilting point is defined as the water content of the soil when most plants will wilt and fail to recover their turgor even if the soil moisture is replenished.

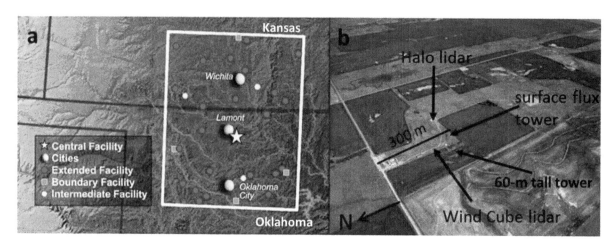

Figure 1. (**a**) Map showing the boundaries of the greater Southern Great Plains (SGP) Atmospheric Radiation Measurement (ARM) Facility in Oklahoma and Kansas (yellow box) including the Central Facility near Lamont, Oklahoma; (**b**) Aerial photograph of the Central Facility depicting the fine-scale spatial heterogeneity in the area and locations of instruments used in this study. Live crops are indicated by dark green while pasturelands and fallow fields are light brown.

We utilized the surface energy flux measurements from the Central Facility's 4 m tall AmeriFlux tower and mean wind speed from a co-located 60 m tall tower. Both towers are well described in [13,31–33]. In brief, the wind vectors are measured on the tall meteorological tower with two 3-D sonic anemometers (Gill-Solent WindMaster Pro, Gill, England) at heights of 25 and 60 m and are available as 30-min averages. The surface flux tower measures carbon dioxide, water vapor, and energy fluxes over an annual cropland with an estimated uncertainty of ±10% based on methods found in [34]. Homogenous fetch is approximately 200 m in length across a 180° arc centered pointing south. The surface flux tower has an open-path infrared gas analyzer (LICOR Li-7500, Li-Cor Biosciences, NE, USA) that measures atmospheric carbon dioxide and water vapor and a 3-D sonic anemometer (Gill-Solent WindMaster Pro) that measures the wind vectors and sonic temperature, from which the sensible heat flux is calculated. From these instruments the Obukhov length (L), a scaling parameter used to indicate atmospheric stability in the surface layer, is also calculated [35]. Other surface meteorology measurements include air temperature, relative humidity, net radiation, soil heat flux, root-zone (0–20 cm) soil moisture, soil temperature and precipitation. The energy fluxes were calculated using a 30-min averaging period and were post-processed using standard algorithms for spike removal, density corrections [36], spectral corrections [37], and coordinate rotation.

It is important to note that the energy flux measurements are taken with separate instruments and full energy budget closure (Rn = H − LE − G − S) can be difficult to achieve. Reasons for this may include different-size instrument footprints, random and systematic measurement errors, incorrect assumptions about instrument time lags for LE (due to sensor separation between the IRGA and sonic anemometer), and uncertainties in spectral corrections (for H and LE). In practice it is generally acceptable for flux towers to achieve only 75%–80% energy closure [38].

The SGP ARM Central Facility has a continuously running Halo scanning lidar (Halo Photonics, Worcestershire, UK) which provides measurements of horizontal wind speed and direction, taken once an hour in 2011, for heights ranging from 75 m to 10 km [39]. In the summer of 2011, the scanning strategy was frequently changed to optimize vertical wind speed retrieval during the middle of the day which further limited the availability of horizontal wind speed data, e.g., 45% of the midday measurements were missing for the second simulation period. Furthermore, the measurements taken close to the surface (<100 m) were deemed suspect due to their location in the near field of the lidar [40]. For these reasons we decided to use the wind speed profile system on the meteorological towers (4, 25, 60 m) in lieu of the lidar for the first two study periods. Wind speeds above the 60 m measurement height were estimated using the power-law expression in Equation (2) with a calculated wind shear exponent (α) from the available data at heights of 25 and 60 m. This was done to estimate the wind speeds at multiple heights above 60 m which represent inflow conditions across the area of a 1.68 MW wind turbine rotor disk (40–120 m).

$$U(z) = U_R (\frac{z}{z_R})^\alpha \qquad (2)$$

In Equation (2) U is the mean horizontal wind speed (m/s) at height z (m), U_R is the mean horizontal wind speed (m/s) at a reference height z_R (m) (here we used the 25 m height), and α is a wind shear exponent used to describe the variations in the wind speed with height. For example, a well-mixed atmosphere will have an α value close to 0 while a strongly stable atmosphere will have α closer to 0.35 [22].

In November 2012, we installed a vertically-profiling Wind Cube v2 pulsed Doppler lidar system (Leosphere, Orsay, France) which measured the u, v and w wind components at 12 vertical levels from 50 m to 180 m above the surface at a ~1 Hz rate and wind speed accuracy of ±0.1 m/s. These data were averaged over 30-min periods to coincide with the energy fluxes. A wind shear exponent was calculated using the 50 m and 80 m heights to best coincide with the methodology used for the meteorological tower. It is important to note that the Wind Cube calculates the wind vectors using a spatial average across a scanned volume cone which increases in diameter with increasing height. This introduces some uncertainty into the measurements as the algorithms used to calculate mean wind speed assume horizontal homogeneity in the sampled flow; although, this source of uncertainty should be relatively low at the flat SGP ARM site. Further uncertainty is introduced by the fact that each measurement level is actually an average of wind flow conditions over a ±10 m probe depth. Even so, studies have shown high agreement between meteorological tower wind speed measurements and those taken by the Wind Cube in flat terrain sites [41,42].

2.2. Case Studies

To study the impact of LSM choice on WRF simulated near-surface wind profiles, we performed atmospheric simulations for three case studies. These cases were associated with a variety of land surface conditions that would affect the surface energy fluxes, including variability in crop cover, albedo, soil moisture, incoming radiation, relative humidity, and air and soil temperature (Table 1, Figure 2).

Table 1. Brief description of Cases 1–3 highlighting canopy and surface meteorology differences.

Variable of Interest	Case 1	Case 2	Case 3
Date	10–24 June 2011	13–27 July 2011	23 November–7 December 2012
Climate characteristics	very warm, wet	hot, dry	cool, dry
Mean air temperature	27.2 °C	32.1 °C	10.2 °C
Precipitation, including the two weeks prior	55.8 mm	8.8 mm	18.1 mm
Mean root-zone soil moisture	24%	6%	12%
Crop	canola	none	none
Field conditions	peak LAI and active canopy in early June then rapid senescence	post-harvest, bare soil with small amounts of crop residue	bare soil with small amounts of crop residue and emerging wheat seedlings

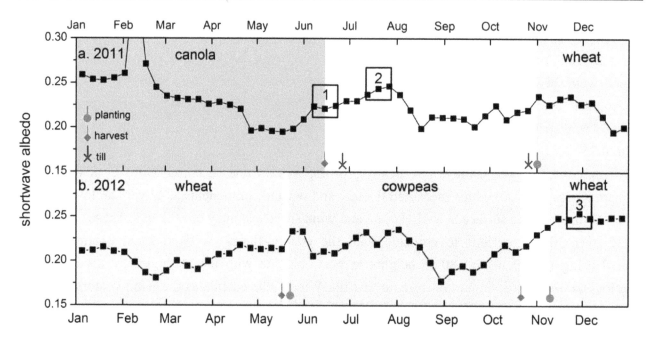

Figure 2. Time series of measured mean weekly albedo at the SGP ARM Central Facility in (**a**) 2011 and (**b**) 2012. Variations in albedo indicate changes in the surface canopy. Each growing season is shaded with a color depending on crop species. Also shown is the timing of planting, harvest and till events. Boxes 1–3 indicate the timing of simulated case periods. Note the differences in albedo between the case studies. The spike in albedo in February 2011 is due to snow cover.

Conditions during Case 1 (10–24 June 2011) included a green canola canopy with peak LAI and adequate access to soil moisture in the root zone ($\theta_v = 24\%$) in early June followed by a rapid crop senescence and crop harvest on 16 June. The field was tilled on 30 June. Although only a month later, Case 2 (13–27 July 2011) conditions included a bare dirt field with some dead crop residue. At this time the field and the surrounding areas were covered with <5% photosynthesizing vegetation and root-zone soil moisture ($\theta_v = 6\%$) was below the wilting point. Case 3 was run to include the autumn (23 November–7 December) of 2012 when the local area was a mixture of <0.5 m tall senesced grasses in the pasturelands and bare ground in the agricultural fields with some dead cowpea residue (<1 cm tall) and a few emerging wheat seedlings (<5 cm tall). Case 3 also included cool temperatures and moderately dry soils (root zone $\theta_v = 12\%$) and was considered a "control" period as energy fluxes would be relatively low during this time of year and should not have a significant influence on PBL behavior and development.

Case 1 and Case 2 (2011) were chosen because they present an interesting test case due to the rapidly observed changes in soil moisture, soil temperature, and atmospheric moisture in comparison to nearly equal incoming radiation. It is not uncommon for such rapid changes to occur in the area due to weather extremes and the type of crop management practices used (e.g., no irrigation). Errors in our WRF runs may be highlighted in such cases since the LSMs did not include high resolution canopy information. Case 3 was chosen because it represents a nice comparison point to Case 2. These two periods differed significantly in meteorology (e.g., air temperature, incoming radiation, importance of synoptic meteorology *versus* local forcings) but both included no canopy.

2.3. WRF Domain Configuration

WRF is a non-hydrostatic, fully compressible atmospheric model that is maintained by the National Center for Atmospheric Research (NCAR) [2]. This study used the advanced research dynamical core version of the WRF 3.4.1 model release with a double nested domain configuration as shown in Figure 3. The outer WRF model domain (domain 1) had a horizontal grid spacing of 9 km and covered a large portion of the central United States. The second domain (domain 2) had a horizontal grid spacing of 3 km while the innermost domain (domain 3) had a grid spacing of 1 km. The model domain configuration was centered on the SGP ARM Central Facility. Given the absence of complex terrain in the innermost model domain, 1 km horizontal resolution was deemed sufficient to resolve fine-scale horizontal flow in the domain of primary interest. A total of 50 terrain-following vertical sigma levels were used for all of the WRF simulations. The 50 sigma level configuration resulted in a vertical resolution of roughly 20–30 meters in the lowest 200 meters of the atmosphere. The top of the model grid was at 50 hPa, which corresponds roughly to a height of 20 km above sea level.

2.4. Ensemble Description

An ensemble consisting of five common LSMs available in WRF 3.4.1 was used to study the impact of LSM choice and configuration on simulating surface energy fluxes and near-surface wind profiles. All runs used the Lin microphysics, Kain-Fritsch cumulus scheme, CAM shortwave and longwave radiation model, Mellor-Yamada-Janjic (MYJ) PBL scheme, and Monin-Obukhov (M-O) surface layer scheme. Surface property input data came from the default 24 USGS land use categories.

In all members, land use was classified as "Dry Cropland and Pastureland". The tested LSMs included: thermal diffusion [43], RUC [44], Pleim-Xiu (PX) [45,46], Noah [47], and Noah Multi-Physics (Noah-MP) [48]. Descriptions of each are provided in Section 2.7.

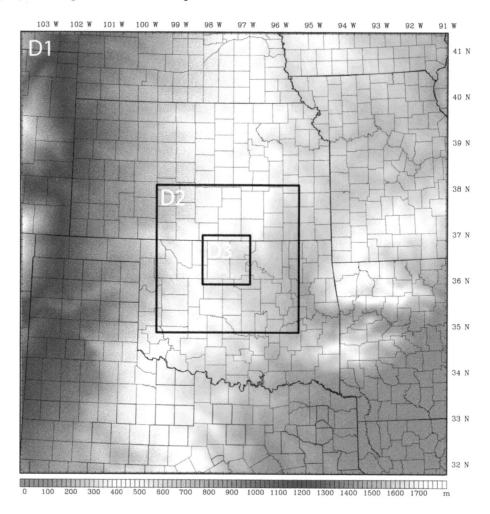

Figure 3. Elevation map showing the WRF model double nested domains used in this study. The inner domain, D3, is centered on the SGP ARM Central Facility in northern Oklahoma, USA. The horizontal grid resolution of domain 1 (labeled D1), domain 2 (D2), and domain 3 (D3) is 9, 3, and 1 km, respectively.

2.5. Input Data

Initial and lateral boundary conditions for the WRF simulations were provided by gridded analysis fields from the North American Model (NAM). The NAM analysis fields are available every 6-h with a horizontal resolution of 12 km. Three-dimensional atmospheric data are provided on 40 pressure levels with a vertical resolution of 25 hPa in the NAM analysis data set. Meteorological variables provided by the NAM data include atmospheric pressure, geopotential height, horizontal wind components, air temperature, relative humidity, surface pressure, sea level pressure, and soil moisture and temperature at four sub-surface layers. NAM analysis data are available for public download from the National Climatic Data Center's (NCDC) data website.

2.6. Four-Dimensional Data Assimilation

WRF has a 4-D data assimilation (FDDA) capability for both analysis (gridded) [49] and observational nudging [50]. The FDDA nudging approach has been demonstrated to reduce error by constraining large-scale atmospheric flow towards analysis fields while allowing smaller scale atmospheric features to develop in fine-scale domains [51–55]. Here we applied analysis nudging on model domains 1 and 2 to constrain the simulated large scale flow of the simulations. Model grid points within the planetary boundary layer were not nudged in either domain 1 and 2 to avoid conflicting with small-scale features that may develop. As an additional constraint to ensure that the nudging algorithm did not influence model results in the lower atmosphere, the FDDA analysis option was turned off below model level 10, which corresponded to a height of roughly 600 m above the surface. The additional FDDA constraint was needed for extremely stable nighttime periods when boundary layer heights of less than 100 m were plausible. Observational nudging was not utilized in any of the WRF domains so that the model solution was free to evolve near the surface where results were later compared with atmospheric observations.

We ran each case study for a spin up period of four days before the WRF output was compared with observations. A multi-day spin up was used to give the soil moisture and temperature initialization-induced LSM biases sufficient time to reach a reasonable balance with underlying atmospheric forcings. All of the model simulations were performed as a single continuous run with no restarts. Output was saved at 10 min intervals for comparison with observations.

2.7. Land Surface Models

Land surface models in WRF use atmospheric information from the surface layer scheme, radiative forcing information from the radiation scheme, and precipitation forcing from the microphysics and convective schemes, together with information about the land surface, to output heat and moisture fluxes. LSMs are described here in order of increasing complexity in regards to how they deal with thermal and moisture fluxes in the soil and vegetation. Additional details are listed in Table 2.

The most basic LSM used here is the thermal diffusion, or "slab", model. Thermal diffusion calculates surface heat fluxes from a 1-dimensional equation and then assumes linear temperature profiles across five layers of soil. Soil moisture is fixed according to land use category and season. Latent energy is computed from the soil moisture availability only. There are no explicit vegetation processes.

The Rapid Update Cycle (RUC) model is largely focused on an accurate characterization of the soil with 6–9 soil levels (or more possible) with the default parameterization reaching down to a soil depth of 300 cm. The physics of snow and phase change in soil are well characterized [44]; however, RUC applies a basic approach to characterize evapotranspiration from the canopy. ET calculations are based on work of [56] wherein the same equations used to calculate the transfers of heat and moisture from the soil are adjusted to estimate the energy fluxes from the canopy. Canopy evapotranspiration is scaled by potential evapotranspiration, canopy moisture content, rooting depth, and the soil moisture relative to the plant wilting point.

Table 2. Configuration of the WRF land surface model (LSM) ensemble. Also listed are the surface runoff and stomatal conductance parameterizations used in Noah-MP. LUC = land use category.

LSM	Vegetation	Drivers of Soil Moisture	Soil Layers	Drivers of Water Flux Exchange
Thermal diffusion	None	LUC	5	Soil surface
RUC	Extension of soil	LUC + evap	6 or 9; 3 m max	Air temperature, relative humidity
Pleim-Xiu (PX)	1-layer from LUC	LUC+ evap+ roots	2; 1 m max	Soil surface + plant transpiration
Noah	1-layer from LUC	LUC+ evap+ roots+ drainage	4; 2 m max	Soil surface + plant transpiration
Noah-MP 1—Ball-Berry/TOPMODEL	2-layer from multiple LUC	LUC + evap + roots + drainage + runoff + storage	Variable; 8 m max	Complex soil–plant feedbacks
Noah-MP 2—Ball-Berry/BATS	2-layer from multiple LUC	LUC + evap + roots + drainage + runoff + storage	Variable; 8 m max	Complex soil–plant feedbacks
Noah-MP 3—Jarvis/TOPMODEL	2-layer from multiple LUC	LUC + evap + roots + drainage + runoff + storage	Variable; 8 m max	Complex soil–plant feedbacks
Noah-MP 4—Jarvis/BATS	2-layer from multiple LUC	LUC + evap + roots + drainage + runoff + storage	Variable; 8 m max	Complex soil–plant feedbacks

The Pleim-Xiu (PX) model [45] builds off of the Noilhan and Planton model [57]. PX only has two soil layers; a shallow layer (1 cm) at the surface interacting with the atmosphere, and a thick layer (99 cm) below interacting with the upper soil layer and the plants via the roots. Moisture fluxes are either from precipitation, dew formation, evaporation directly from the soil surface or vegetation, transpiration (ultimately controlled by soil moisture in the root zone via stomatal resistance), or vertical movement within the soil. Regardless of canopy cover or density, only a soil heat capacity is used to calculate soil heat flux.

The Noah model [47] is the most complex of the LSMs evaluated. Noah is the product of a large community of researchers combining previous iterations of multiple land surface models. The model has four soil layers reaching 2 m in which soil temperature and moisture are calculated. The first three layers form the root zone in non-forested areas. The shallow soil, "leaky" bottom (precipitation that infiltrates to the bottom of the soil column is lost from the system) and simple snow/melt/thaw dynamics are seen as areas where the model avoids complexity. Vegetation, however, is well defined using monthly estimates of albedo and fraction of green vegetation cover based on vegetation class. Vertical rooting depth for crops is also allowed to change from month to month. Evapotranspiration can be modeled by the Ball-Berry equation, taking both the physics of water flow through the soil and plants as well as the physiology of photosynthesis into account. Alternatively, a simpler Jarvis scheme based only on leaf area index and physics of water flow through the soil–vegetation–atmosphere continuum can be used. These advancements were implemented to improve parameterizations related to seasonal and diurnal simulations of water fluxes, energy fluxes and state variables [58].

Multiple physics parameterization options are available with the Noah model (Noah-MP) for users to tune the model in order to maximize complexity for dominant processes, while using simpler modeling approaches for other processes [48]. For example, Noah-MP includes three options for modeling the response of transpiration to changes in site water status: via soil moisture directly as in Noah and via two variations of the response of stomatal conductance to soil matric potential. Site hydrology can be modeled simply as in Noah with the "leaky" bottom either with the simplified calculations of surface runoff or with more complex calculation of surface runoff based on the BATS model [59]. Alternatively, the TOPMODEL-based [60] calculation of surface runoff and groundwater discharge can be used with varying levels of complexity in modeling groundwater. Here we ran four versions of Noah-MP, including two versions of a surface runoff model (BATS, TOPMODEL) and two versions of canopy stomatal resistance model (Ball-Berry, Jarvis) to assess any differences in simulated surface energy fluxes due to these parameterizations.

3. Results and Discussion

3.1. Land-Atmosphere Energy Exchange

3.1.1. Observations

Case 1 and Case 2 were chosen as they represented very different surface conditions across a relatively short time scale (e.g., <month) during a period with similar incoming radiation. The rapid changes in crop cover, albedo, and surface meteorology are evident in the soil moisture and energy flux measurements, the latter shown by the midday Bowen ratio (Figure 4). Average daily soil moisture during Case 1 was 24% in comparison to 6% in Case 2. Likewise, significant differences in midday β were apparent as the measured Bowen ratio in Case 2 was ten times higher ($\beta = 10$) than during Case 1 ($\beta = 0.9$). Mean midday latent energy fluxes reached just 50 W·m^{-2} during Case 2 leading to very high β values while the LE flux reached 250 W·m^{-2} in Case 1. These differences in LE are largely explained by the presence of a transpiring canopy, periods of evaporation from a wet canopy or wet ground, and soil moisture above the wilting point during much of Case 1 while Case 2 was characterized by post-harvest conditions including dry, bare ground. During Case 2, 75% of the available energy in the afternoon was re-emitted to the atmosphere in the form of sensible heat. In Case 3, we saw much lower H and LE fluxes due to less available net radiation during the autumn months. Average soil moisture during Case 3 was 12% and the midday average β was 4.5.

The storage of energy in the crop canopy is negligible at this site; however, storage of energy in the ground surface (G) is significant at times. During daylight hours in Case 1, G compromised 8% of available energy while in Cases 2 and 3, G was a larger portion at 13% due to the lack of canopy cover. Average midday energy budget closure was 83% in Case 1, 96% in Case 2, and 96% in Case 3. The "missing" or unaccounted energy during Case 1 is likely due to an underestimation of the LE flux. LE is more difficult to accurately measure than H, and LE was a significant component of the net energy balance during Case 1.

3.1.2. LSM Performance

The LSM runs were initialized with NAM soil moisture values close to observed values in Case 1 and Case 3 but with wetter than actual conditions in Case 2. In fact, the NAM data showed no dry down between Case 1 and 2 as the initialized soil moisture conditions were the same for both periods ($\theta_v = 15\%$). We also compared the LSM simulated soil moisture on days 4 and 14 of each simulation for all models except for thermal diffusion, as this LSM does not predict soil moisture. In Case 1, a couple of the LSMs (RUC, PX) overestimated the drying out of the soil over the two week period and simulated soil moisture values equal to 6%–7% by the end of the two weeks. This occurred even as significant precipitation fell during week 2. In contrast, average soil moisture simulated by the Noah-MP runs was 20%–24% at the end of the simulations and in agreement with the observations. In Case 2, most of the LSMs underestimated the drying out of the soil partially because the initial conditions were so much higher than those observed. For example, Noah and Noah-MP simulated soil moisture values (mean $\theta_v = 14\%$) above the wilting point ($\theta_v = 10\%$) at the end of the period. Only two of the LSMs had declining soil moisture in agreement with the observations: RUC ($\theta_v = 7\%$) and PX ($\theta_v = 6\%$). In Case 3 there was very little difference in soil moisture between the LSMs by the end of the simulation period as the atmospheric conditions (low radiation, low temperature) limit evaporative water loss. The simulated θ_v values were in agreement with the observations for Case 3 for all LSMs.

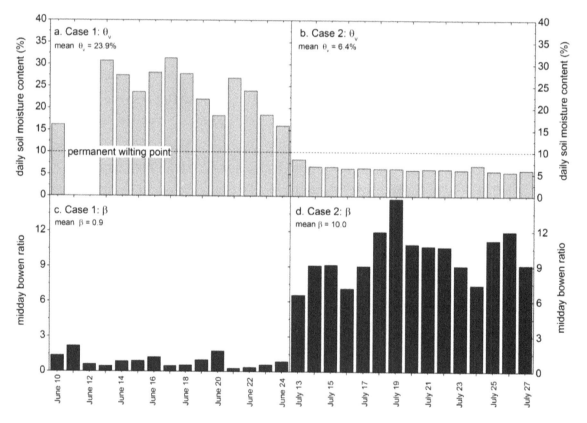

Figure 4. Mean daily soil moisture content measurements in (**a**) June 2011/Case 1 and (**b**) July 2011/Case 2. Mean midday Bowen ratio (β) measurements from the surface flux tower are also plotted for (**c**) June 2011/Case 1 and (**d**) July 2011/Case 2. The permanent wilting point is for a well-drained Kirkland soil.

Mean measured and simulated Rn, H, LE and G fluxes are shown in Figures 5–7 by time of day for each case period. Although Rn and G are not discussed in detail they are shown in the diurnal plots to illustrate how well each model simulated the full energy budget. Mean midday Bowen ratios are listed in Table 3 for each case period.

Table 3. Mean (± standard deviation) midday Bowen ratios by case period show differences across LSMs and seasons.

Observations/LSM	Case 1	Case 2	Case 3
Observations	0.92 ± 0.62	9.90 ± 4.52	5.33 ± 2.30
Thermal diffusion	0.20 ± 0.12	0.23 ± 0.11	0.76 ± 0.27
RUC	4.43 ± 4.15	132.45 ± 237.36	2.47 ± 2.43
PX	1.12 ± 0.36	1.36 ± 1.02	1.29 ± 0.53
Noah	1.82 ± 0.26	4.40 ± 0.82	2.54 ± 1.58
Noah-MP 1	1.55 ± 0.27	5.89 ± 2.37	9.01 ± 3.91
Noah-MP 2	2.05 ± 0.38	5.47 ± 2.61	8.64 ± 3.98
Noah-MP 3	1.97 ± 0.39	7.55 ± 3.34	8.34 ± 3.84
Noah-MP 4	1.12 ± 0.38	7.41 ± 2.92	8.41 ± 3.86
LSM Range	0.20–4.43	0.23–132.45	0.76–9.01

For Case 1, mean midday β ranged from 0.2 (thermal diffusion) to 4.4 (RUC) while most simulations fell between $\beta = 1$–2, just slightly higher than the mean observed value ($\beta = 0.9$) (Table 3). Nearly all of the LSMs in Case 1 overestimated midday H fluxes (Figure 5). The greatest overestimation was by RUC as this model predicted a mean midday H equal to 500 W·m^{-2} while the mean observation peaked at 150 W·m^{-2}. For comparison, RUC also simulated much drier soil moisture conditions than the observations. Although thermal diffusion was relatively accurate in predicting the buoyant heat flux, this model, which does not include plant dynamics or predictions of soil moisture, greatly overestimated LE which led to a very low Bowen ratio.

Small but noticeable differences in mean midday β were apparent in the Noah-MP runs as Noah-MP 4 predicted the lowest Bowen ratio ($\beta = 1.1$) of the Noah-MP group and Noah-MP 2 simulated the highest ($\beta = 2.1$). These models differed in the choice of stomatal conductance model but were run with the same soil hydrology model (BATS) (Table 2). Noah-MP 3 was slightly lower than Noah-MP 2, however not significantly, indicating that the choice of soil hydrology model also played a role. This suggests that LE fluxes in June were driven by evaporation fluxes in addition to transpiration as there were periods of precipitation during this case. The soil hydrology models used here in Noah-MP differ in the way they parameterize surface runoff, which was especially important during the second half of Case 1. Simulated soil moisture for the Noah MP runs was similar during the beginning of the simulations but deviated by the end (θ_v ranged from 18% to 22%). Prior studies have shown that differences in surface runoff characterization can be significant, for example, the study of [61] found that the TOPMODEL led to significant improvements in simulating surface runoff for sites in the Mississippi River Basin.

The results for Case 2 are found in Figure 6. Large errors were found in simulated H and LE fluxes for thermal diffusion (overestimated LE and underestimated H), in H for RUC (overestimated), and in LE for PX (overestimated). Exact reasons for these errors are unknown especially since RUC and PX were the most accurate of all the LSMs in simulating surface soil moisture. These errors are evident in the large range of simulated mean midday Bowen ratios: 0.2 (thermal diffusion) and 132 (RUC). The observed mean midday β was 10 (Table 3).

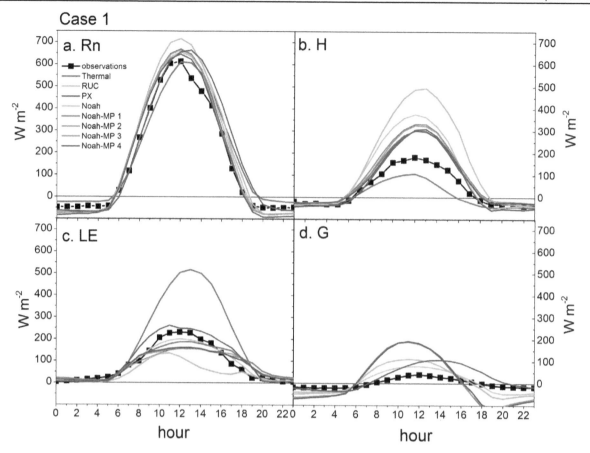

Figure 5. Mean diurnal plots of measured and simulated **(a)** net available energy (Rn), **(b)** sensible heat (H), **(c)** latent energy (LE), and **(d)** ground heat (G) fluxes for each LSM during Case 1 (June 2011). Positive H and LE fluxes indicate net energy transfer to the atmosphere. Positive G fluxes indicate net energy transfer to the ground surface. Time is given in Local Standard Time.

Significant differences within the Noah-MP models were also apparent in Case 2. The Bowen ratios were similar between Noah-MP 1 and 2 ($\beta \sim 5.5$) and Noah-MP 3 and 4 ($\beta \sim 7.5$). These groups are divided by the parameterization of stomatal conductance whereby Noah-MP 1-2 used the more complex Ball-Berry equation. Noah-MP runs 1 and 2 simulated slightly wetter soil moisture values than Noah-MP 3 and 4. Since these Bowen ratio differences are divided by the choice of stomatal conductance parameterization it appears that the vegetative indices or LAI values used as input into the LSMs were not reflective of actual conditions as no vegetation was growing during this period at the site. Hence, the type of model used to parameterize photosynthesis and thus transpiration should have had no effect on the simulated Bowen ratios if the models had the correct vegetation cover information as input.

Noah and Noah-MP were the most accurate LSMs during the no-canopy, low soil moisture periods even as the actual plant–atmosphere feedbacks were negligible and these models overestimated soil moisture. We had expected RUC to perform well during Case 2 as this model has sophisticated soil layers and drivers that are based on atmospheric temperature and humidity instead of physiologically driven controls. RUC, however, overestimated H and predicted low LE fluxes leading to unrealistically high midday Bowen ratios (>100).

Case 2

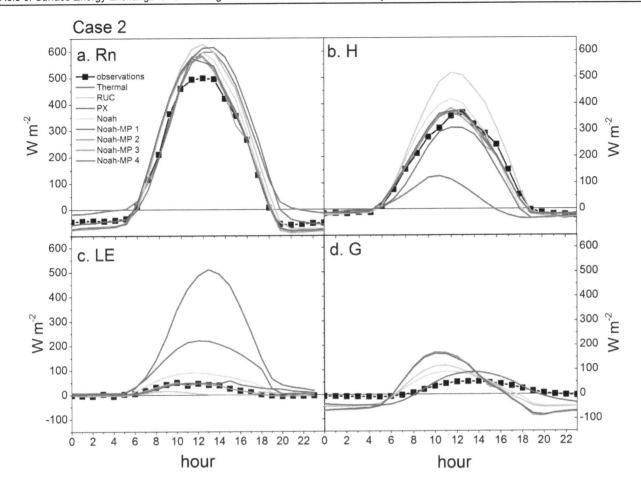

Figure 6. Mean diurnal plots of measured and simulated **(a)** net available energy (Rn), **(b)** sensible heat (H), **(c)** latent energy (LE), and **(d)** ground heat (G) fluxes for each LSM during Case 2 (July 2011).

Both the simulated and observed energy fluxes during Case 3 were much lower in autumn than the summer months due to lower incoming radiation (Figure 7). The largest errors occurred for simulated sensible heat as the thermal diffusion, RUC and PX models all underestimated the midday daytime buoyant heat flux. Thermal diffusion and PX also largely overestimated the LE flux. Mean midday Bowen ratios for these two models were 0.76 and 1.3, respectively, while the observed mean midday β was 5.3 (Table 3). During Case 3 we saw the largest differences between Bowen ratios for Noah and Noah-MP whereby mean midday β was 2.5 for Noah and ranged from 8.3 to 9.0 for the Noah-MP group. However, even small LE and H differences, as was the case during Case 3, can create large Bowen ratio differences due to less available energy in the system. This is evident in the wide range of β magnitudes simulated by the LSMs (Table 3). Despite this instability in the Bowen ratio, Noah and Noah-MP both performed reasonably well in simulating the energy budget.

While the surface property and meteorology conditions were very different amongst the case periods, some of the LSMs produced a very narrow range of Bowen ratios. Mean midday β ranged from 0.2 (Case 1) to 0.8 (Case 3) for thermal diffusion, which had no plant dynamics and uses a fixed value for soil moisture, and from 1.1 (Case 1) to 1.4 (Case 2) for PX. In comparison, the range was 0.9 (Case 1) to 10 (Case 2) for the flux observations. LSMs with similar variance included the Noah-MP group which ranged from 1.5 (Case 1) to 9 (Case 3) although the highest Bowen ratio was found for the third case and not the second as we saw in the observations. It is interesting that the PX model had such a small range of Bowen ratios amongst the case studies since vegetation processes are relatively complex in this model.

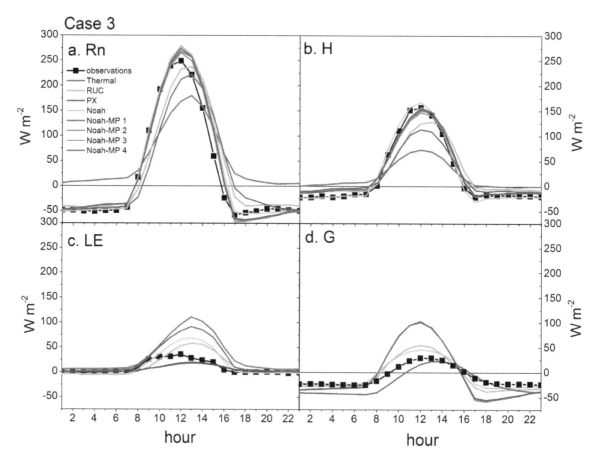

Figure 7. Mean diurnal plots of measured and simulated **(a)** net available energy (Rn), **(b)** sensible heat (H), **(c)** latent energy (LE), and **(d)** ground heat (G) fluxes for each LSM during Case 3 (November-December 2012).

Vegetation plays a key role in the Noah-MP model. For example, stomatal conductance determines rates of photosynthesis and transpiration, the dynamic leaf model predicts leaf area index and green vegetation fraction, and the canopy water scheme simulates canopy water interception and evaporation. Even so, we saw little to no significant improvement of Noah-MP over Noah. We expected to see largest differences between Noah and Noah-MP as well as variability within the Noah-MP runs during Case 1 when the vegetation-atmosphere feedbacks were most important. However, the simulation results were not consistent with this theory. Work from other studies may explain this finding. While Noah-MP has shown improvement in natural ecosystems (forests, grasslands), no significant improvement over baseline Noah has been noted for Noah-MP simulations of LE in croplands as both models tend to overestimate LE fluxes for croplands [61,62]. This is likely due to the fact that neither the leaf dynamics in Noah-MP nor the monthly LAI estimates used by Noah can capture managed crop growth [61]. Simulating crops in Noah may require a dynamic leaf and root growth module to enhance performance as found in [63].

3.2. Rotor-Disk Wind Shear

Due to the change in surface meteorology and canopy characteristics from June to July 2011, Cases 1 and 2 had very different energy partitioning schemes (Figures 4–6). As such, we expected to see less wind shear in Case 2 due to stronger daytime mixing as the buoyancy heat flux, or sensible heat flux, dominated the surface energy exchange. Atmospheric mixing or stability was described by the Obukhov length (L). Cases 1 and 2, as expected, showed differences in atmospheric mixing during the midday hours. Average midday L was −128 m during Case 1 and L = −12 m during Case 2. These magnitudes correspond to forced convective conditions in Case 1 and strongly convective in Case 2 [22,64].

Differences between Case 1 and Case 2 were also apparent in the measured mean midday wind shear exponent where α was 0.12 (higher shear) in June and 0.02 (lower shear) in July across heights equal to a wind turbine rotor disk. In June the wind speed at the top of the rotor disk was over 1 m/s faster than at the bottom of the rotor disk. Shear across the rotor disk not only changes the magnitude of the inflow or the rotor disk-equivalent wind speed (e.g., [20,22,65]), but can also put fatigue loads on turbine components which reduce turbine performance and lifetimes [66]. This is especially true if the shear is related to intense turbulence bursts (*i.e.*, strong coherent structures) such as those experienced during a nocturnal low level jet (LLJ). At sites without LLJs, on the other hand, shear across the rotor-disk may increase the average wind speed in the inflow conditions (*i.e.*, rotor disk wind speed > hub-height wind speed) leading to greater power generation than otherwise is produced during a well-mixed environment (*i.e.*, rotor disk wind speed = hub-height wind speed) [23].

Wind speed profiles for the measurements as well as the simulations in all three cases are presented in Figure 8. Here, the wind speed at each height is plotted as the difference between the measured or simulated speed at height z and the measured or simulated speed at height 80 m. This was done to ease comparisons between the simulated and observed profiles, however, the shape of the profiles in these plots do not correspond directly to wind shear exponent magnitudes. For example, wind shear appears to be far greater during Case 1 than Case 3, however this is largely a function of higher wind speeds at all heights during Case 1 as the mean midday α values between each cases were nearly identical.

Simulated wind shear exponents ranged from 0.08 (RUC) to 0.15 (thermal diffusion) in June, 0.05 (RUC) to 0.10 (thermal diffusion) in July, and 0.09 (Noah-MP) to 0.15 (PX) in November–December in comparison to the observations of α = 0.12, 0.02, and 0.13 in those periods, respectively. For perspective, a study performed at a West Coast wind farm found that α values < 0 correspond to strongly convective conditions, 0 < α < 0.1 correspond to convective conditions, and 0.1 < α < 0.2 correspond to neutral conditions [22]. Near-neutral conditions in June may have been caused by frequent periods of high wind speeds, cloud cover and a weakened buoyant heat flux.

In Case 1, all of the models under-predicted wind shear in the top half of the turbine rotor disk, leading to underestimations of the total rotor disk wind speed (Figure 8a). In this case, underestimations of rotor disk wind speed from the models could lead to under-predictions of power produced by the wind turbines. Both the models and the observations indicated convective conditions in July although the models slightly under-predicted the strength of the midday mixing with the highest error occurring in the thermal model (Figure 8b). Greatest error is seen here in the lower half of the rotor-disk as the models over-estimated wind shear between 40 and 80 m a.g.l. as compared to the observations. Wind shear exponent magnitudes indicate near-neutral stability in Case 3. Agreement between the models and observations during the autumn period were overall high for the top half of the rotor-disk, with greatest accuracy occurring in the Noah and Noah-MP model runs (Figure 8c).

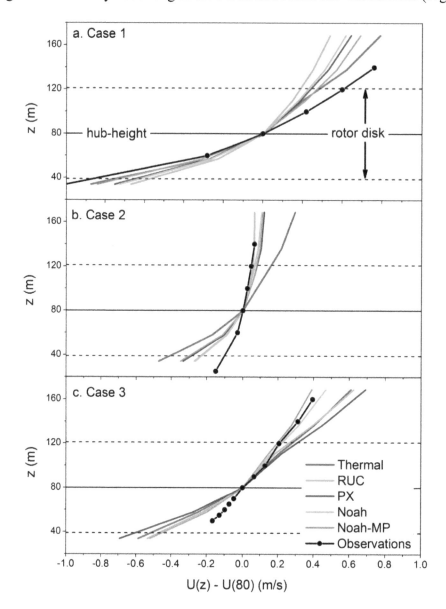

Figure 8. Mean midday wind speed profiles during **(a)** Case 1, **(b)** Case 2, and **(c)** Case 3 for each LSM in comparison to the observations shown in black. The four Noah-MP runs are shown as a group average. The wind speeds are plotted as the difference between the wind speed at height (z) and the wind speed at 80 m (wind turbine hub-height). Blades on nearby turbines span from 40 to 120 m in the atmosphere. Differences in shear profiles represent differences in the mean wind speed across a turbine rotor-disk.

3.3. Relationship between Wind Shear and Surface Energy Exchange

The following analysis was done to identify periods of significant land surface-atmosphere feedbacks. A relationship between midday wind shear and midday Bowen ratio in June and July is seen in the observations. Cases 1 and 2 in Figure 9 show a decline in wind shear with increasing β; *i.e.*, increased atmospheric mixing is associated with a higher portion of the available energy re-emitted as the buoyant heat flux. Slopes were −0.019 in June and −0.006 in July. As expected this trend was not seen in November–December (slope = +0.003) namely because land surface-atmosphere feedbacks are much weaker during the autumn months at this site.

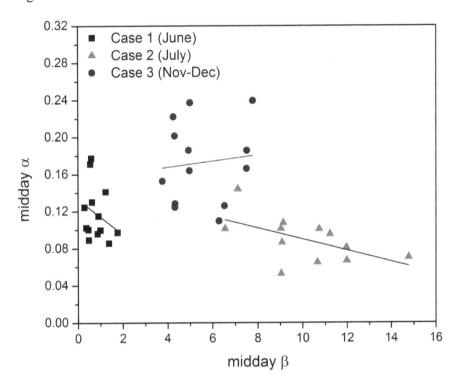

Figure 9. Scatter plots of observed wind shear exponents (α) and observed Bowen ratios (β) for all three cases. Plotted are daily mean midday values. This plot shows a similar correlation between wind shear and surface energy exchange during the summer months. In Cases 1 and 2 higher β (sensible heat > latent heat) was correlated with less wind shear (lower α). This relationship is not seen during the cool, dry Case 3 period which indicates a weaker surface-atmosphere forcing in November–December.

Figure 10 shows the relationship between mean midday Bowen ratio and wind shear parameter for each LSM in comparison to the mean observation for each case period. As in Figure 9, Figure 10 also shows a relationship between simulated energy flux partitioning and simulated wind shear. For example, simulations of greater wind shear (higher α) were correlated to lower Bowen ratios, such as those predicted by the thermal diffusion and PX models. Simulations of lower wind shear or a well-mixed atmosphere were correlated to higher Bowen ratio predictions, such as RUC in Case 1 and Noah-MP in Case 3. One exception was the RUC performance in Case 2 whereby RUC simulated an outlier Bowen ratio that did not fit the pattern.

In Case 1 the LSMs closest to the observations included those with explicit vegetation-atmosphere processes (PX, Noah and Noah-MP); however, the complexity of these feedbacks (such as those found in Noah-MP) did not improve model performance. In Case 2 nearly all of the models predicted more shear and lower Bowen ratios than the observations with the exception of RUC which largely overestimated β. In Case 3 the LSMs predicted a wide range of Bowen ratios which is partially attributed to the sensitivity of β to small H and LE magnitudes (*i.e.*, available energy was much lower during the autumn than during the summer months). These plots suggest a strong connection between the magnitude of simulated wind shear and the partitioning of surface energy fluxes in the model. These results also suggest that LSM choice in WRF via land-atmosphere connections has a significant impact on simulated wind flow at heights relevant to wind energy.

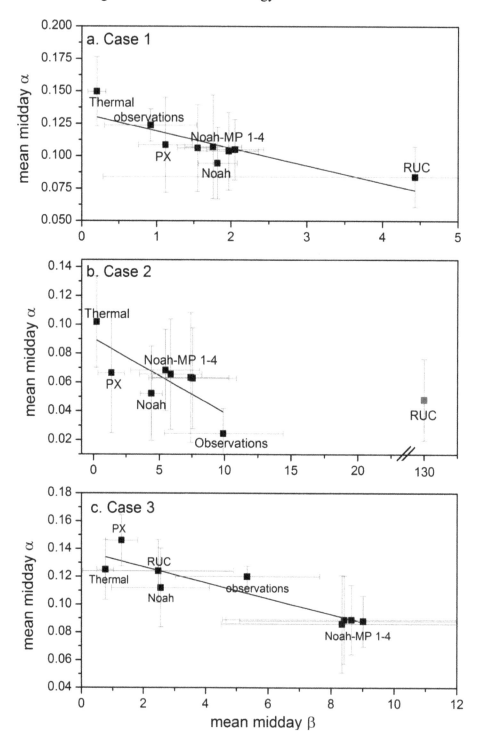

Figure 10. Scatter plot of mean (± standard deviation) midday values of Bowen ratio and wind shear exponent for each LSM and the observations for (**a**) Case 1 (June 2011); (**b**) Case 2 (July 2011) and (**c**) Case 3 (November–December 2012). The data are fitted with a linear regression model (Case 1 slope = −0.013, R^2 = 0.66; Case 2 slope = −0.005, R^2 = 0.55; Case 3 slope = −0.006, R^2 = 0.82). Models which simulated higher Bowen ratios (sensible heat > latent heat) tended to simulate less wind shear.

4. Conclusions

We conducted this study to look for relationships between observed wind shear and land surface energy exchange processes and to determine whether the choice of land surface model in WRF significantly influences the magnitudes of simulated wind inflow at rotor-disk height.

The flux tower and wind profile observations showed a relationship between energy partitioning at the surface and wind shear during the warm summer months when surface-atmosphere energy exchange coupling should be at its peak. As expected, stronger convection and less wind shear occurred during periods when buoyant heat fluxes dominated the surface energy exchange. This relationship was not found during the autumn "control" period; however, as surface energy forcing was relatively weak.

Overall, the simplest LSMs tended to have the poorest performance in estimating the surface energy budget and to a smaller extent with simulating wind shear. These models included the thermal diffusion model and the RUC model; however, the latter model had very inconsistent results across the three cases. These two models have either simple or no vegetative-atmosphere feedback processes. In addition, thermal diffusion uses a fixed soil moisture value based only on land use category. Making further conclusions based on model sophistication of soil–plant–atmosphere processes is however difficult as the PX, Noah, and Noah-MP performances were not consistently different from each other.

We had expected to see seasonal differences in model performance as the observed canopy and surface meteorological conditions varied greatly across the three cases. However, the range of LSM errors did not appear to differ dramatically from one period to another. Between Case 1 and Case 2, it is likely that the models are responding to atmospheric drivers (e.g., precip, temp, wind speed) from the boundary conditions as the land surface conditions (e.g., land-use category, z_0) did not significantly change in the models. For example, in Case 1, Noah-MP predicted a higher Bowen ratio than observed (*i.e.*, higher sensible heat flux, lower ET flux) possibly indicating that the model's classification of the canopy was not as photosynthetically-active as observed in the field. This may be because canopy values are averaged across the whole life-span of crops whereas in Case 1 the crops were (for a time at least) at peak density and peak ET rates. While in Case 2, Noah-MP is using the same land surface conditions as in Case 1 and underestimates the Bowen ratio whereas the observations show higher Bowen ratios as is expected for bare soil.

Case 2 (summer) and Case 3 (winter) were dominated by differences in atmospheric drivers for the surface energy fluxes, e.g., temperature and precipitation, in addition to differences in the model's interpretation of surface conditions. Despite Cases 2 and 3 having the same surface type (bare soil), the models that consider vegetation are interpreting both cases as having a crop canopy with varying surface roughness.

Generally we saw larger simulation errors in the wind speed profiles below hub-height (<80 m) than at heights above. This error affects the accuracy of the simulated rotor-disk wind speed. For example, in Cases 2 and 3 the simulated wind speeds in the bottom half of the rotor-disk were less than those observed. We also saw differences in the wind shear parameters amongst the LSM members even during the cool autumn Case 3 when land surface energy-atmosphere feedbacks should be negligible. This is because the wind speed profile is sensitive to other factors including the surface drag induced by canopy roughness. For example, in [16] the authors found that differences in the roughness length, z_0, from changing the landscape from a corn canopy (z_0 = 0.25 m) to soybean (z_0 = 0.10 m) in the Noah LSM, increased hub-height wind speed and decreased wind shear in the U.S. Midwest. As changes in surface roughness affect surface drag, they also saw larger effects on wind shear in the bottom half of the rotor-disk. Although our study did not explicitly consider the impact of z_0 on the simulated wind profiles, we did see noticeable differences in z_0 of up to 0.10 m both across seasons and amongst LSMs. The largest differences between models occurred during Case 3 whereby values for PX (z_0 = 0.12 m) were higher than those found for thermal diffusion and RUC (z_0 = 0.05 m). The largest differences between seasons occurred within both thermal diffusion and RUC whereby summer values were 0.10 m greater than those in winter. These differences in z_0 may partially explain why we saw differences in wind shear during the winter "control" run in Case 3.

A portion of our simulation errors are also due to a mismatch between spatial and temporal scales. We ran WRF down to 1 km (mesoscale) resolution while the flux tower and wind profile instruments had a footprint closer to 200 m. Flow could have significant variability even across these scales with the existing sharp gradients of land use and soil moisture in this area. WRF was run with a land use category defined as Dry Cropland and Pastureland while in reality the canopy type varied across space and time during the simulation periods. The LSMs used here with the finest temporally adjusted land cover input were Noah and Noah-MP. A monthly time step may not have high enough precision. For example, the crop rapidly changed during two of our cases, Case 1 and 2. The crop quickly senesced over a period of a couple of weeks but this information was not in the standard versions of WRF. For example, WRF was reinitialized with new soil moisture conditions between the June and July 2011 simulations but it used the same land use based surface characteristics even though in reality the field was tilled during these four weeks in 2011. Further improvements to simulating the wind fields in this area could be done by using land use data with finer spatial and temporal resolution. For example, [33] found that the SGP region requires the characterization of vegetation properties in LSMs at a realistic spatial scale for accurately estimating surface energy fluxes. Thus, the SGP region may need LSMs with the capability of accurately specifying surface characteristics on a sub-grid scale, such as from satellite-derived products, to capture fine scale variability, including senescence, harvest and tillage events. Whether data assimilation of surface properties would also improve wind forecasting was beyond the scope of this paper but warrants exploration.

This study was conducted as an evaluation of running each LSM in WRF in "standard" mode as a wind farm operator might do without significantly fine-tuning each LSM to ensure the best performance. For example, some of the LSMs investigated here have options for remote sensing surface data assimilation (e.g., RUC, Noah-MP) which may have enhanced model performance. With these caveats in mind we showed that wind shear is sensitive to the choice of LSM. This is a significant finding, as wind shear across the rotor-disk has a significant impact on the power produced by the wind turbine [20,23]. Any improvement in short-term wind forecasting should drive down the cost of wind energy and further wind as a leading form of low cost renewable energy in the coming decade [67].

Lastly, we did not have available data to illustrate a link between surface energy exchange processes and turbine power production via the influence on the wind shear profiles. Although the SGP ARM site is located a few miles east of some of the turbines within a multi-MW wind farm, turbine measurements of power were not available during the case studies modeled here as the wind farm did not begin operations until 2013. New simulations should be done during periods of concurrent wind power data to assess the relationship between surface energy exchange, wind shear and power production at this wind farm.

Acknowledgments

This work was funded by the US Department of Energy, Wind and Water Power Technologies Office under agreement #WBS 02.07.00.01. The flux tower and surface meteorology observations were supported by the Office of Biological and Environmental Research of the US Department of Energy under contract No. DE-AC02-05CH11231 as part of the Atmospheric Radiation Measurement (ARM) and Atmospheric System Research programs. The authors would like to thank the staff at the SGP ARM Central Facility for their help with maintaining the Wind Cube lidar system and their support throughout our study. We'd also like to thank Marc Fischer and Margaret Torn (LBNL) for their support with the flux tower data and Rob Newsom (PNNL) for his support with the Halo lidar data. LLNL is operated by Lawrence Livermore National Security, LLC, for the DOE, National Nuclear Security Administration under contract DE-AC52-07NA27344. LLNL-JRNL-658967

Author Contributions

Sonia Wharton was responsible for project management and led the project design and data analysis. Matthew Simpson and Jessica Osuna ran the WRF simulations and provided discussion on the LSMs. Jennifer Newman helped with the lidar deployment and data analysis. Sebastien Biraud provided the surface flux and meteorological tower data and provided guidance on those datasets. All authors contributed to the writing of this manuscript.

Conflicts of Interest

The authors declare no conflict of interest.

References

1.   Lew, D.; Milligan, M.; Jordan, G.; Piwko, R. *Value of Wind Power Forecasting*; NREL Report No. CP-5500–50814; NREL: Golden, CO, USA, 2011.
2.   Skamarock, W.C.; Klemp, J.B.; Dudhia, J.; Gill, D.O.; Barker, D.M.; Duda, M.; Huang, X.-Y.; Wang, W.; Powers, J.G. A Description of the Advanced Research WRF Version 3. NCAR Technical Report 2008. Available online: http://www.mmm.ucar.edu/wrf/users/docs/arw_v3.pdf (accessed on 2 February 2014).
3.   American Wind Energy Association. Oklahoma Wind Energy Fact Sheet April 2014. Available online: http://awea.files.cms-plus.com/images/oklahoma.pdf (accessed on 21 May 2014).
4.   Wiser, R.; Bolinger, M.; Barbose, G.; Darghouth, N.; Hoen, B.; Mills, A.; Weaver, S.; Porter, K.; Buckley, M.; Fink, S.; Oteri, F.; Tegen, S. *2012 Wind Technologies Market Report*; DOE/GO-102013–3948; U.S. Department of Energy, Office of Energy Efficiency and Renewable Energy: Washington, DC, USA, 2013.

5. Chen, F.; Bornstein, R.; Grimmond, S.; Li, J.; Liang, X.; Martilli, A.; Miao, S.; Voogt, J.; Wang, Y. Research priorities in observing and modeling urban weather and climate. *Bull. Am. Meteorol. Soc.* **2012**, *93*, 1725–1728.

6. Anderson, D.E.; Verma, S.B. Carbon-dioxide, water-vapor and sensible heat exchanges of a grain-sorghum canopy. *Bound. Layer Meteor.* **1986**, *34*, 317–331.

7. Baldocchi, D.D. A comparative study of mass and energy exchange rates over a closed C-3 (wheat) and an open C-4 (corn) crop. Part 2: CO_2 exchange and water use efficiency. *Agric. For. Meteorol.* **1994**, *67*, 291–321.

8. Betts, A.K.; Ball, J.H.; Beljaars, A.C.M.; Miller, M.J.; Viterbo, P.A. The land surface-atmosphere interaction: A review based on observational and global model perspectives. *J. Geophys. Res.* **1996**, *101*, 7209–7225.

9. Raupach, M.R. Equilibrium evaporation and the convective boundary layer. *Bound. Layer Meteor.* **2000**, *96*, 107–141.

10. Patton, E.G.; Sullivan, P.P.; Moeng, C.-H. The influence of idealized heterogeneity on wet and dry planetary boundary layers coupled to the land surface. *J. Atmos. Sci.* **2004**, *62*, 2078–2097.

11. Desai, A.R.; Davis, K.J.; Senff, C.J.; Ismail, S.; Browell, E.V.; Stauffer, D.R.; Reen, B.P. A case study of the effects of heterogeneous soil moisture on mesoscale boundary-layer structure in the Southern Great Plains, U.S.A. Part 1: Simple prognostic model. *Bound. Layer Meteorol.* **2006**, *119*, 195–238.

12. Maxwell, R.M.; Chow, F.K.; Kollet, S.J. The groundwater-land surface-atmosphere connection: Soil moisture effects on the atmospheric boundary layer in fully-coupled simulations. *Adv. Water Resourc.* **2007**, *30*, 2447–2466.

13. Fischer, M.L.; Billesbach, D.P.; Riley, W.J.; Berry, J.A.; Torn, M.S. Spatiotemporal variations in growing season exchanges of CO_2, H_2O, and sensible heat in agricultural fields of the Southern Great Plains. *Earth Interact.* **2007**, *11*, 1–21.

14. Storm, B.; Basu, S. The WRF model forecast-derived low-level wind shear climatology over the United States Great Plains. *Energies* **2010**, *3*, 258–276.

15. Draxl, C.; Hahmann, A.N.; Peña, A.; Giebel, G. Evaluating winds and vertical wind shear from Weather Research and Forecasting model forecasts using seven planetary boundary layer schemes. *Wind Energy* **2014**, *17*, 39–55.

16. Vanderwende, B.J.; Lundquist, J.K. Could crop height impact the wind resource at agriculturally productive wind farms? In Proceedings of the 21st Symposium on Boundary Layers and Turbulence, Leeds, UK, 9–13 June 2014.

17. Rohatgi, J.; Barbezier, G. Wind turbulence and atmospheric stability—Their effects on wind turbine output. *Renew. Energy* **1999**, *16*, 908–911.

18. Sumner, J.; Masson, C. Influence of atmospheric stability on wind turbine power performance curves. *J. Sol. Energy Eng.* **2006**, *128*, 531–537.

19. Van den Berg, G.P. Wind turbine power and sound in relation to atmospheric stability. *Wind Energy* **2008**, *11*, 151–169.

20. Wagner, R.; Antoniou, I.; Pedersen, S.M.; Courtney, M.S.; Jørgensen, H.E. The influence of the wind speed profile on wind turbine performance measurements. *Wind Energy* **2009**, *12*, 348–362.

21. Vanderwende, B.J.; Lundquist, J.K. The modification of wind turbine performance by statistically distinct atmospheric regimes. *Environ. Res. Lett.* **2012**, doi:10.1088/1748-9326/7/3/034035.

22. Wharton, S.; Lundquist, J.K. Assessing atmospheric stability and its impact on rotor disk wind characteristics at an onshore wind farm. *Wind Energy* **2012**, *15*, 525–546.

23. Wharton, S.; Lundquist, J.K. Atmospheric stability affects wind turbine power collection. *Environ. Res. Lett.* **2012**, doi:10.1088/1748-9326/7/1/014005.

24. Trier, S.B.; Davis, C.A.; Ahijevych, D. Environmental controls on the simulated diurnal cycle of warm-season precipitation in the Continental United States. *J Atmos. Sci.* **2010**, doi:10.1175/2009JAS3247.1.

25. LeMone, M.A.; Grossman, R.L.; Coulter, R.L.; Wesley, M.L.; Klazura, G.E.; Poulos, G.S.; Blumen, W.; Lundquist, J.K.; Cuenca, R.H.; Kelly, S.F.; *et al.* Land–atmosphere interaction research and opportunities in the Walnut River watershed in Southeast Kansas: CASES and ABLE. *Bull. Am. Meteorol. Soc.* **2000**, *81*, 757–779.

26. Weckwerth, T.; Parsons, D.B.; Koch, S.E.; Moore, J.A.; LeMone, M.A.; Demoz, B.B.; Flamant, C.; Geerts, B.; Wang, J.; Feltz, W.F. An overview of the International H_2O Project (IHOP_2002) and some preliminary highlights. *Bull. Am. Meteorol. Soc.* **2004**, *85*, 253–277.

27. LeMone, M.A.; Chen, F.; Alfieri, J.G.; Tewari, M.; Geerts, B.; Miao, Q.; Grossman, R.L.; Coulter, R.L. Influence of land cover and soil moisture on the horizontal distribution of sensible and latent heat fluxes in Southeast Kansas during IHOP_2 and CASES-97. *J. Hydrometeorol.* **2007**, *8*, 68–87.

28. Miller, J.; Barlage, M.; Zeng, X.; Wei, H.; Mitchell, K.; Tarpley, D. Sensitivity of the NCEP/Noah land surface model to the MODIS green vegetation fraction data set. *Geophys. Res. Lett.* **2006**, doi:10.1029/2006GL026636.

29. Case, J.L.; LaFontaine, F.J.; Kumar, S.V.; Peters-Lidard, C.D. Using the NASA-Unified WRF to assess the impacts of real-time vegetation on simulations of severe weather. In Proceedings of the 13th Annual WRF User's Workshop, Boulder, CO, USA, 26–29 June 2012; pp. 1–69.

30. Torn, M.S.; Biraud, S.C.; Still, C.J.; Riley, W.J.; Berry, J.A. Seasonal and interannual variability in 13C composition of ecosystem carbon fluxes in the U.S. Southern Great Plains. *Tellus* **2011**, *63B*, 181–195.

31. Billesbach, D.P.; Fischer, M.L.; Torn, M.S.; Berry, J.A. A portable eddy covariance system for the measurement of ecosystem-atmosphere exchange of CO_2, water vapor, and energy. *J Atmos. Ocean. Technol.* **2004**, *21*, 639–650.

32. Riley, W.J.; Biraud, S.C.; Torn, M.S.; Fischer, M.L.; Billesbach, D.P.; Berry, J.A. Regional CO_2 and latent heat surface fluxes in the Southern Great Plains: Measurements, modeling, and scaling. *J. Geophys. Res.* **2009**, doi:10.1029/2009JG001003.

33. Raz-Yaseef, N.; Fischer, M.L.; Billesbach, D.P.; Biraud, S.C.; Gunter, S.A.; Bradford, J.A.; Torn, M.S. Drought sensitivity of three U.S. southern plains ecosystems. *Environ. Res. Lett.* submitted.

34. Finkelstein, P.L.; Sims, P.F. Sampling error in eddy correlation flux measurements. *J. Geophys. Res.* **2001**, *106*, 3503–3509.

35. Obukhov, A.M. Turbulence in an atmosphere with a non-uniform temperature. *Bound. Layer Meteorol.* **1971**, *2*, 7–29.

36. Webb, E.K.; Pearman, G.I.; Leuning, R. Correction of flux measurements for density effects due to heat and water vapour transfer. *Quart. J. Roy. Meteorol. Soc.* **1980**, *106*, 85–100.

37. Moore, C.J. Frequency response corrections for eddy correlation systems. *Bound. Layer Meteorol.* **1986**, *37*, 17–35.

38. Wilson, K.; Goldstein, A.; Falge, E.; Aubinet, M.; Baldocchi, D.; Berbigier, P.; Bernhofer, C.; Ceulemans, R.; Dolman, H.; Field, C.; *et al.* Energy balance closure at FLUXNET sites. *Agric. For. Meteorol.* **2002**, *113*, 223–243.

39. Newsom, R.K. Doppler Lidar Handbook. Office of Biological and Environmental Research, Office of Science, U.S. Department of Energy: Washington, DC, USA. Available online: http://www.arm.gov/publications/tech_reports/handbooks/dl_handbook.pdf (accessed on 15 December 2014)

40. Emeis, S. *Measurement Methods in Atmospheric Sciences: In-Situ and Remote*; Borntraeger Science Publishers: Stuttgart, Germany, 2010.

41. Betancur, S.; Lemoine, R.; Boinne, R.; Mail, P.-A.; Dewilde, L. Reducing risk in resource assessment using lidar. In Proceedings of the WindExpo Conference, Guadalajara, Mexico, 5–7 November 2008.

42. Cañadillas, B.; Westerhellweg, A.; Neumann, T. Testing the performance of a ground-based wind LiDAR system: One year intercomparison at the offshore platform FINO1. *Dewi Mag.* **2011**, *38*, 58–64.

43. Blackadar, A.K. Modeling pollutant transfer during daytime convection. In Proceedings of the Fourth Symposium on Atmospheric Turbulence, Diffusion and Air Quality, Reno, NV, USA, 15–18 January 1979; pp. 443–447.

44. Smirnova, T.G.; Brown, J.M.; Benjamin, S.G. Performance of different soil model configurations in simulating ground surface temperature and surface fluxes. *Mon. Weather Rev.* **1997**, *125*, 1870–1884.

45. Pleim, J.E.; Xiu, A. Development and testing of a surface flux and planetary boundary layer model for application in mesoscale models. *J. Appl. Meteor.* **1995**, *34*, 16–32.

46. Xiu, A.; Pleim, J.E. Development of a land surface model. Part I: Application in a mesoscale meteorological model. *J. Appl. Meteor.* **2001**, *40*, 192–209.

47. Chen, F.; Mitchell, K.; Schaake, J.; Xue, Y.; Pan, H.-L.; Koren, V.; Duan, Q.Y.; Ek, M.; Betts, A. Modeling of land surface evaporation by four schemes and comparison with FIFE observations. *J. Geophys. Res.* **1996**, *101*, 7251–7268.

48. Niu, G.-Y.; Yang, Z.-L.; Mitchell, K.E.; Chen, F.; Ek, M.B.; Barlage, M.; Kumar, A.; Manning, K.; Niyogi, D.; Rosero, E.; *et al.* The community Noah land surface model with multiparameterization options (Noah-MP): 1. Model description and evaluation with local-scale measurements. *J. Geophys. Res.* **2011**, doi:10.1029/2010JD015139.

49. Stauffer, D.R.; Seaman, N.L. On multi-scale four-dimensional data assimilation. *J. Appl. Meteor.* **1994**, *33*, 416–434.

50. Liu, Y.; Bourgeois, A.; Warner, T.; Swerdlin, S.; Hacker, J. An implementation of obs-nudging-based FDDA into WRF for supporting ATEC test operations. In Proceedings of the WRF User Workshop, Boulder, CO, USA, 27–30 June 2005; Paper 10.7.

51. Stauffer, D.R.; Seaman, N.L. Use of four-dimensional data assimilation in a limited-area mesoscale model. Part I: Experiments with synoptic-scale data. *Mon. Weather Rev.* **1990**, *118*, 1250–1277.

52. Lo, J.C.F.; Yang, Z.L.; Pielke, R.A., Sr. Assessment of three dynamical climate downscaling methods using the Weather Research and Forecasting (WRF) model. *J. Geophys. Res.* **2008**, doi:10.1029/2007JD009216.

53. Otte, T.L. The impact of nudging in the meteorological model for retrospective air quality simulations. Part I: Evaluation against national observation networks. *J. Appl. Meteorol. Climatol.* **2008**, *47*, 1853–1867.

54. Salathé, E.P.; Steed, R.; Mass, C.F.; Zahn, P.H. A high-resolution climate model for the U.S. Pacific Northwest: Mesoscale feedbacks and local responses to climate change. *J. Clim.* **2008**, *21*, 5708–5726.

55. Bowden, J.H.; Otte, T.L.; Nolte, C.G.; Otte, M.J. Examining interior grid nudging techniques using two-way nesting in the WRF model for regional climate modeling. *J. Clim.* **2012**, *25*, 2805–2823.

56. Pan, H.-L.; Mahrt, L. Interaction between soil hydrology and boundary layer development. *Bound. Layer Meteorol.* **1987**, *38*, 185–202.

57. Noilhan, J.; Planton, S. A simple parameterization of land surface processes for meteorological models. *Mon. Weather Rev.* **1989**, *117*, 536–549.

58. Xia, Y.; Mitchell, K.; Ek, M.; Sheffield, J.; Cosgrove, B.; Wood, E.; Luo, L.; Alonge, C.; Wei, H.; Meng, J.; *et al.* Continental-scale water and energy flux analysis and validation for the North American Land Data Assimilation System project phase 2(NLDAS-2): 1. Intercomparison and application of model products. *J. Geophys. Res.* **2012**, doi:10.1029/2011JD016048.

59. Yang, Z.L.; Dickinson, R.E. Description of the Biosphere-Atmosphere Transfer Scheme (BATS) for the Soil Moisture Workshop and evaluation of its performance. *Glob. Planet. Chang.* **1996**, *13*, 117–134.

60. Niu, G.-Y.; Yang, Z.-L.; Dickinson, R.E.; Gulden, L.E. A simple TOPMODEL-based runoff parameterization (SIMTOP) for use in global climate models. *J. Geophys. Res.* **2005**, doi:10.1029/2005JD006111.

61. Cai, X.; Yang, Z.-L.; David, C.H.; Niu, G.-Y.; Rodell, M. Hydrological evaluation of the Noah-MP land surface model for the Mississippi River basin. *J. Geophys. Res.* **2014**, doi:10.1002/2013JD020792.

62. Blyth, E.; Gash, J.; Lloyd, A.; Pryor, M.; Weedon, G.P.; Shuttleworth, J. Evaluating the JULES land surface model energy fluxes using FLUXNET data. *J. Hydrometeorol.* **2010**, *11*, 509–519.

63. Gaylor, S.; Wohling, T.; Grzeschik, M.; Ingwersen, J.; Wizemann, H.D.; Warrach-Sagi, K.; Hogy, P.; Attinger, S.; Streck, T.; Wulfmeyer, R. Incorporating dynamic root growth enhances the performance of Noah-MP at two contrasting winter wheat field sites. *Water Resourc. Res.* **2014**, *50*, 1337–1356.

64. Van Wijk, A.J.M.; Beljaars, A.C.M.; Holtslag, A.A.M.; Turkenburg, W.C. Evaluation of stability corrections in wind speed profiles over the North Sea. *J. Wind Eng. Ind. Aerodyn.* **1990**, *33*, 551–566.

65. Elliott, D.L.; Cadogan, J.B. Effects of wind shear and turbulence on wind turbine power curves. In Proceedings of the European Community Wind Energy Conference and Exhibition, Madrid, Spain, 10–14 September 1990.

66. Kelley, N.; Shirazi, M.; Jager, D.; Wilde, S.; Adams, J.; Buhl, M.; Sullivan, P.; Patton, E. *Lamar Low-Level Jet Project Interim Report*; NREL/TP-500–34593; NREL: Golden, CO, USA, 2004; pp. 1–216.

67. Department of Energy, Energy Efficiency and Renewable Energy. *Wind Program Complex Flow Workshop Report*; Department of Energy, Energy Efficiency and Renewable Energy: Boulder, CO, USA, 2012; pp. 1–122.

9

Basic Concepts for Convection Parameterization in Weather Forecast and Climate Models: COST Action ES0905 Final Report

Jun–Ichi Yano [1,*], **Jean-François Geleyn** [1], **Martin Köhler** [2], **Dmitrii Mironov** [2], **Johannes Quaas** [3], **Pedro M. M. Soares** [4], **Vaughan T. J. Phillips** [5], **Robert S. Plant** [6], **Anna Deluca** [7], **Pascal Marquet** [1], **Lukrecia Stulic** [8] and **Zeljka Fuchs** [8]

[1] CNRM/GAME UMR 3589, Météo-France and CNRS, 31057 Toulouse Cedex, France;
E-Mails: mma109@chmi.cz (J.-F.G.); pascal.marquet@meteo.fr (P.M.)

[2] DWD, 63067 Offenbach, Germany; E-Mails: Martin.Koehler@dwd.de (M.K.);
Dmitrii.Mironov@dwd.de (D.M.)

[3] Institute for Meteorology, Universität Leipzig, 04103 Leipzig, Germany;
E-Mail: johannes.quaas@uni-leipzig.de

[4] Instituto Dom Luiz, Faculdade de Ciências, Universidade de Lisboa, 1749-016 Lisboa, Portugal;
E-Mail: pmsoares@fc.ul.pt

[5] Department of Physical Geography,and Ecosystem Science, University of Lund, Solvegatan 12, Lund
223 62, Sweden; E-Mail: vaughan.phillips@nateko.lu.se

[6] Department of Meteorology, University of Reading, RG6 6BB Reading, UK;
E-Mail: r.s.plant@reading.ac.uk

[7] Max Planck Institute for the Physics of Complex Systems, 01187 Dresden, Germany;
E-Mail: adeluca@pks.mpg.de

[8] Department of Physics, University of Split, 21000 Split, Croatia;
E-Mails: lukrezsia@gmail.com (L.S.); zeljka.fuchs@gmail.com (Z.F.)

* Author to whom correspondence should be addressed; E-Mail: jiy.gfder@gmail.com

Academic Editor: Katja Friedrich

Abstract: The research network "Basic Concepts for Convection Parameterization in Weather Forecast and Climate Models" was organized with European funding (COST Action ES0905) for the period of 2010–2014. Its extensive brainstorming suggests how the subgrid-scale parameterization problem in atmospheric modeling, especially for convection, can be examined and developed from the point of view of a robust theoretical basis. Our main cautions are current emphasis on massive observational data analyses and process studies. The closure and the entrainment–detrainment problems are identified as the two highest priorities for convection parameterization under the mass–flux formulation. The need for a drastic change of the current European research culture as concerns policies and funding in order not to further deplete the visions of the European researchers focusing on those basic issues is emphasized.

Keywords: parameterization; convection; subgrid scales

---

1. Introduction

The research network COST Action ES0905 "Basic Concepts for Convection Parameterization in Weather Forecast and Climate Models" was organized with European funding over the period 2010–2014.

The present paper constitutes a final scientific report of the network activity. The achievements of the present Action are closely examined by following each task agreed, and each question listed in the Memorandum of Understanding (MoU, available at: www.cost.eu/domains_actions/ essem/Actions/ES0905). In some cases, a question in concern turns out to be ill-posed or ambiguous. The present report acknowledges such instances, and indeed, considers the need for a re-evaluation of some aspects of the MoU to be an outcome of the Action in its own right. The report is only concerned with the scientific developments and achievements of the Action network. For information on the actual Action activities (meetings and documents), the readers are strongly encouraged to visit the Action Web site (http://convection.zmaw.de/).

Our major achievement as a deliverable is a monograph [1]. Reference to this monograph is made frequently throughout the report for this reason. However, the present report is not an abbreviation of the monograph: it is another key deliverable that, by complementing the monograph, more critically analyzes the current convection parameterization issues and presents the future perspectives.

The present report is assembled by the lead author under his responsibility as a chair of the present Action. Many have contributed to this specific process. Those who have critically contributed are asked to be co-authors, including working group (WG) leaders (JFG, MK, DM, JQ). Many others have contributed by submitting text segments, by proof reading part of or the entirety of the text, as well as general comments. These contributions are listed in the acknowledgments at the end of the present report. Though the present report does not intend to replace an official final report submitted to the COST office, it expresses collective positions of the contributing authors while also reflecting well the overall Action achievements.

By its nature, the present report does not intend to be a comprehensive review, but focus more specifically on the issues listed in MoU and actually discussed in due course of the COST Action. The report particularly does not always remark on the issues already accepted to be important in the community. The emphasis is, more than often, on the issues that are not widely appreciated, and the report tends to take an unconventional view in order to compensate for the currently widely accepted views. For this reason, the report is overall, more critical with the current state-of-the-art of the research. However, we request that the readers not to read us as critics for the sake of critics. The contributing authors also bear these criticisms as our common fault of failing to do better than otherwise. The intention of these critical remarks are so that our convection parameterization research becomes more sound in future to come. Of course, the authors do not claim that all the arguments developed in the present report are ultimately right, but are only the best ones that they can offer for now. Thus, these arguments must be used as a starting point for more critical and constructive debates on the future direction of the convection parameterization research. The general perspectives in the last section of the report are also developed from this point of view.

1.1. Overview

The main objective of the Action has been, as stated in MoU, "to provide clear theoretical guidance on convection parameterization for climate and numerical weather prediction models". Here, the problem of *parameterization* arises because both the weather–forecast and the climate models run only with limited spatial resolutions, and thus many physical processes are not properly represented by falling short of the resolution required for adequate explicit simulation of the process. In other words, there are processes in the "subgrid scales", which must somehow be included as a part of a model in an indirect, parameteric manner. Such a procedure is called *parameterization* (*cf.* [2]). Convection is one of the key processes to be parameterized considering its importance in heat and moisture budget of the atmosphere (*cf.* [3]), and the one upon which the present Action is focused.

Here, although the word "parameterization" is often used in a much wider sense in the literature, in the present report, the term is strictly limited to the description of subgrid–scale processes. Many of the atmospheric physical processes (notably cloud microphysics) must often be *phenomenologically* described by vast simplifications. However, such a phenomenological description should not be confused with the parameterization problem.

The parameterization problem is often considered a highly technical, "engineering" issue without theoretical basis. Often, it is even simply reduced to a matter of "tuning". The goal of the Action is to suggest how the parameterization problem can be addressed from a more basic theoretical basis, both from perspectives of theoretical physics and applied mathematics. For this purpose, the extensive brain storming has been performed by organizing a number of meetings. Here, the results of these theoretical reflections are reported by closely following what has been promised in the MoU. The focus is on the technical questions listed in MoU. However, a more basic intention is, by examining these questions, to suggest more generally what can be addressed from more fundamental perspectives of theoretical physics and applied mathematics, and how. For this reason, efforts are always made to add general introductory remarks in introducing each subject.

1.2. A Key Achievement

MoU specifically lists (Sec. B.1 Background) the following convection-related processes that were still to be resolved: too early onset of afternoon convection over land, underestimation of rainfall maximum, failure to represent the 20–60 day planetary–scale tropical oscillation (the Madden–Julian oscillation). We can safely claim that one of the problems listed there, the afternoon convection, is now solved by the efforts under the present Action. As it turns out, the key is to examine a closure in convection parameterization in a more careful manner, as will be further discussed in **T1.1**. This is also considered a good example case for demonstrating the importance of theoretical guidance on the convection parameterization problem.

1.3. Identified Pathways

As MoU states, "The Action proposes a clear pathway for more coherent and effective parameterization by integrating existing operational schemes and new theoretical ideas". As it turns out, instead of a unique pathway, we identify three major pathways for pursuing such an endeavor [4]: (1) fundamental turbulence research; (2) close investigations of the parameterization formulation itself; and (3) better understanding of the processes going on both within convection and the boundary layer.

The first approach is based on our understanding that the atmospheric convective processes are fundamentally turbulent. Thus, without fundamental understanding of the latter, no real breakthrough can be expected in the convection parameterization problem. The second is rather conventional, and even often considered obsolete, but we very strongly emphasize the importance of a thorough understanding how a given parameterization actually works in order to improve it. We believe that proper emphasis on the first two major pathways is critically missing in the current research efforts. The second is probably even more important than the first for the reasons to be discussed below.

The third is the currently most widely-accepted approach with the use of cloud–resolving models (CRMs) as well as large–eddy simulations (LESs). We are rather critical of the current over-reliance on this approach. Our criticisms are double edged: these studies must be performed with a clear pathway leading to an improvement of a parameterization in mind. At the same time, much in–depth analysis for identifying precise mechanisms (e.g., under energy cycle: *cf.* Q1.2.1) associated with a given process is required (*cf.* Section 2.4.1).

Here, the second approach should not be confused with a more conventional, "blind" tuning. Emphasized here is an importance of in-depth understanding of a parameterization *itself* in order to improve it, and we have to know why it must be modified in a particular manner, and we should also be able to explain why the model can be improved in this manner. Such a careful parameterization study should not be confused with the process studies, either.

A good historical lesson to learn from turbulence research is an improvement of the so-called QN (quasi–normal) model into the EDQN (eddy–damping quasi–normal) model in simulating the turbulent kinetic energy spectrum (*cf.* Ch. VII in [5]). This improvement was not achieved by any process study of turbulent motions, but rather a close investigation of the QN model itself, even at a level of physical variables which are not explicitly evaluated in actual simulations. As identified, the skewness, one of such variables, tends to steepen with time in a rather singular manner. Thus it suggests an additional damping to the skewness equation is necessary. This further suggests how the kinetic energy equation must be modified consistently.

This is not a simple "tuning" exercise, but a real improvement based on a physical understanding of the behavior of a given parameterization. Without in–depth understanding, no real improvement of a parameterization would be possible. A process study has no contribution here either. Just imagine, if Orszag [6] has had focused on intensive process studies of turbulence based on, say, direct numerical simulations (DNSs): he would never have identified a problem with the skewness equation in QN. Note that the key issue is in self–consistency (cf. **T2.4** below) of the QN formulation itself, but nothing to do with number of physical processes incorporated into QN.

1.4. Model Comparisons and Process Studies

We emphasize, as stated in MoU, that the aim of the present Action is to "complement" the already existing model comparison studies. In other words, from the outset, we did not intend to perform model comparison studies by ourselves *a priori*. Rather we are skeptical against those existing model comparison studies, especially for development and verification of parameterizations.

Of course, this is not to discredit all the benefits associated with model comparison exercises. A single–column configuration typically adopted for comparison studies is extremely helpful for identifying the workings of subgrid–scale parameterizations in a stand–alone manner with large–scale (resolved–scale) processes prescribed as column–averaged tendencies. These tendencies are often taken from observations from a field campaign, thus these are expected to be more reliable than those found in stand–alone model simulations, or even a typical assimilation data for forecast initialization. In this manner, the role of individual subgrid–scale processes can clearly be examined. For example, the surface fluxes can be prescribed so that the other processes can be examined without feedback from the surface fluxes. Merits of running several models together are hardly denied either. In this way, we can clearly see common traits in a certain class of parameterization more clearly than otherwise (e.g., [7,8]).

On the other hand, a process study associated with a model comparison can lead to misleading emphasis for the parameterization development. For example, Guichard *et al.* [7] suggested the importance of transformation from shallow to deep convection in a diurnal convective cycle, as phenomenologically inferred from their CRM simulations. No careful investigation on a possible mechanism behind was performed there. This has, nevertheless, led to extensive process studies on the shallow–to–deep transition (e.g., [9] and the references therein: see also **Q1.2.1**, **Q2.3.1**). This conclusion would have been certainly legitimate for guiding a direction for the further studies of the convective processes. However, Guichard *et al.* [7] even suggested this transformation process as a key missing element in parameterizations. The difference between these two statements must clearly be distinguished. A suggested focus on the transformation process does not give any key where to look within a parameterization itself: is it an issue of entrainment–detrainment or closure (cf. Section 2.1)?

The lead author would strongly argue that this work has led to a rather misguided convection–parameterization research over the following decade. In short sight, it is easier to control convection height by entrainment and detrainment rather than by closure. Thus, vast research effort is diverted to the former with an overall neglect of the latter.

As it turns out, the closure is rather a key issue for this problem as demonstrated by Bechtold *et al.* [10] and fully discussed in **T1.1**, but rather in contrast to other earlier attempts. Here, one may argue that the transformation process is ultimately linked to the closure problem. However, how can we see this simply by many process experiments? In addition, how we identify a possible modification of the closure in this manner? It would be similar to asking to Orszag [6] to run many DNSs and figure out a problem in the skewness equation of QN. Note that even a skewness budget analysis of DNSs would not point out the problem in QN. We should clearly distinguish between process studies and the parameterization studies: the latter does not follow *automatically* from the former.

1.5. Organization of the Action

In the next section, we examine our major achievements by following our four major activities:

1. Mass-flux based approaches (Section 2.1)
2. Non-Mass Flux based approaches (Section 2.2)
3. High-Resolution Limit (Section 2.3)
4. Physics and Observations (Section 2.4)

These four activities, respectively roughly cover the four secondary objectives listed in MoU (Section C.2):

1. Critical analysis of the strengths and weakness of the state-of-the-art convection parameterizations
2. Development of conceptual models of atmospheric convection by exploiting methodologies from theoretical physics and applied mathematics
3. Proposal of a generalized parameterization scheme applicable to all conceivable states of the atmosphere
4. Defining suitable validation methods for convection parameterization against explicit modeling (CRM and LES) as well as against observations, especially satellite data

These secondary objectives are furthermore associated with a list of Tasks to be achieved and a list of Questions to be answered in MoU. In the following sections, we examine how far we have achieved the promised Tasks, and then present our answers for the Questions listed for each category in the MoU. Note that these Tasks and Questions are given by bold–face headings starting with the upper–case initials, T and Q, respectively.

The assigned numbers in MoU are used for the Tasks, whereas the MoU does not assign any numbers to the Questions. Here, the order of the Questions is altered from the MoU so that they are presented side–by–side with the listed Tasks in order. Numbers are assigned to the Questions accordingly. On the other hand, the order and the numbering of the Tasks are unaltered from MoU, which is considered like a legal document binding us. This choice is made in order to make it clear that the present report is written directly in response to MoU, though the pre-given order may not be the best. To compensate, extensive inter-references between the Tasks and Questions are given in the text.

In answering these tasks and questions, various new theoretical ideas are often outlined. However, we consider that full development of these ideas are beyond the scope of the present report. References to other recent papers produced by the Action members are made when appropriate to allow readers to delve more thoroughly into some of those ideas that have been pursued to date.

2. Tasks and Questions

2.1. Mass-Flux Based Approaches

The majority of both operational weather–forecast and climate–projection models adopt mass–flux based approaches for convection parameterization.

In mass–flux based approaches, the key issues clearly remain the closure and the entrainment-detrainment [11]. In spite of progress under the present Action, we are still short of identifying ultimate answers to both issues. Thus, the best recommendation we can make is to re–emphasize an importance of focusing on these two key issues in future research on convection parameterization so long as we decide to stay with the mass–flux based formulation. Presently, maintenance of mass–flux formulation is the basic strategy of all the major operational research centers. For this reason, the following discussion is also naturally focused on these two issues along with other related issues.

However, it should be recognized that the mass–flux formulation is not without limit. Remember especially that this formulation is specifically designed to represent "plume" type convection such as convective cumulus towers as well as smaller–scale equivalent entities found in the boundary–layer and over the inversion layer (cloud topped or not). The formulation clearly does not apply to disorganized turbulent flows typically found in the boundary layer on much smaller scales. An important distinction here is that transport by convective plumes and turbulent mixing (background turbulence) are inherently non–local and local, respectively. Thus, a qualitatively different description is required. The last point may be important to bear in mind because these disorganized flows are likely to become more important processes to be parameterized with increasing horizontal resolutions of the models (*cf.* Section 2.3).

2.1.1. Overview

Reminders of some basics of mass-flux convection parameterization (*cf.* [12,13] are due first. As the name suggests, the quantity called mass flux, M, which measures vertical mass transport by convection, is a key variable to be determined. Once it is known, in principle, various remaining calculations are relatively straightforward in order to obtain the final answer of the grid-box averaged feedback of convection, as required for any subgrid-scale parameterization. This approach works well so long as we stay with a standard thermodynamics formulation (with the standard approximations: *cf.* [14–16], **T2.4** below) and microphysical processes (including precipitation) can be neglected. The latter must either be drastically simplified in order to make it fit into the above standard formulation, or alternatively, an explicit treatment of convective vertical velocity is required (*cf.* [17]: see further **Q2.1.2** below). The last is a hard task by itself under the mass-flux formulation, as reviewed in a book chapter [18].

As assumed in many operational schemes, the mass flux can be separated into two factors, one for the vertical profile and the other for a time-dependent amplitude:

$$M = \eta(z)M_B(t) \tag{1}$$

Here, a subscript B is added to the amplitude $M_B(t)$, because customarily it is defined at convection base, although it is misleading to literally consider it to be determined at the convection base for the reason to be explained immediately below (see also **Q1.3.1**).

Such a separation of variables becomes possible by assuming a steady state for parameterized convective ensembles. This assumption is usually called the "steady plume" hypothesis because these convective ensembles are usually approximated by certain types of plumes. This assumption naturally comes out when the convective scale is much smaller than that of the "resolved" large scales. Under this situation, the time scale for convection is so short that we may assume that convective ensembles are simply in equilibrium with a large-scale state.

Under this hypothesis, we should not think in terms of a naive picture that convection is initiated from a boundary-layer top and gradually grows upwards. Such a transient process is simply not considered [19]. As a whole, under this standard approximation, a life cycle of individual convective clouds, including an initial trigger, is not at all taken into account. In other words, in order to include those processes, this approximation must first be relaxed.

Under the standard "plume" formulation, a vertical profile, $\eta(z)$, of mass flux is determined by the entrainment and the detrainment rates, E and D:

$$\frac{\partial M}{\partial z} = E - D \tag{2}$$

Here, entrainment and detrainment, respectively, refer to influx and outflux of air mass into and out of the convection (convective plume), as the above formula suggests. Thus, a key issue reduces to that of prescribing the entrainment and the detrainment rates. Once these parameters are known, a vertical profile of mass flux, $\eta(z)$, can be determined in a straightforward manner, by vertically integrating Equation (2: *cf.* **T1.4**). Here, however, note a subtle point that strictly the mass–flux profile, $\eta(z)$, also changes with time through the change of the entrainment and the detrainment rates by following the change of a large–scale state.

A standard hypothesis (convective quasi-equilibrium hypothesis) is to assume that convection is under equilibrium with a given large-scale state. As a result, the amplitude of convection is expected to be determined solely in terms of a large-scale state. This problem is called the closure.

Thus, closure and entrainment-detrainment are identified as the two key problems. Here, it is important to emphasize that, against a common belief, the trigger is not a part of the mass-flux convection parameterization problem for the reason just explained. Though we may choose to set the convective amplitude to zero (because convection does not exist always), this would simply be a part of the closure formulation.

T1.1: Review of Current State-of-the-Art of Closure Hypothesis

Our review on the closure [20] has facilitated in resolving the afternoon convection problem [10]: the question of onset of convection in late afternoon rather than in early afternoon, as found globally over land, by following the sun with the maximum of conditional instability as conventionally measured by CAPE (convective available potential energy). Many efforts were invested on modifying entrainment–detrainment parameters (e.g., [21,22]), because they appeared to control the transformation from shallow to deep convection (*cf.* Section 1.4). However, as it turns out, the key is rather to improve the closure. Note that the modifications of entrainment–detrainment also achieves this goal, but typically in expense of deteriorating the model climatology. Our effort for a systematic investigation on the closure problem [20] has greatly contributed in identifying this key issue. Bechtold *et al.* [10], in turn, have actually implemented our key conclusion into operation.

The closure strategy tends to be divided into two dichotomous approaches by strongly emphasizing the processes either in the boundary layer (boundary-layer controlled closure [23]) or in the free troposphere (parcel-environment based closure [24,25]). The boundary-layer controlled closure tends to be more popular in the literature (e.g., [26–28]), probably due to the fact that the boundary layer is rich with many processes, apparently providing more possibilities. However, as clearly pointed out by Donner and Phillips [29], the boundary-layer control closure does not work in practice for mesoscale convective systems, which evolve slowly over many hours, because the processes in the boundary layer are too noisy to be useful as a closure condition in forecast models. Though some global climate models do adopt boundary-layer controlled closures [30–32], their behavior tends to noisier than otherwise (*cf.* Figures 3, 4, 6 and 7 of [32], respectively). Under the parcel–environment based closure, it is rather the large–scale forcing (e.g., uplifting) from the free troposphere that controls convection. Recall that the trigger from the boundary layer is *not* a part of the standard mass–flux formulation, as already remarked. Though a process study on trigger of convection over a heterogeneous terrain, for example, may be fascinating (*cf.* [33]), no direct link with the closure problem should be made prematurely (*cf.* Section 2.4.1.2).

The basic idea of the parcel-environment based closure is to turn off the influence of the boundary layer for modifying CAPE with time when constructing the closure. In this manner, an evolution of parameterized convection not influenced by noisy boundary–layer processes is obtained. The review by Yano *et al.* emphasizes the superiority of the parcel-environment based closure against the boundary–layer controlled closure [20]. After its completion, this parcel-environment based closure is actually adopted at ECMWF. It is found that this relatively straightforward modification of the closure essentially solves the problem of the afternoon convection (a proper phase for the convective diurnal cycle) without any additional modifications to the model [10]. This implementation does not rely on any tuning exercise either: the result is rather insensitive to the major free parameter T^* (in their own notation) for the range of 1–6 K against the reported choice of $T^* = 1$ K [34]. The reported model improvement does not depend on details of the entrainment-detrainment, either [35].

Nevertheless, we should not insist that the former closure always works, or that the boundary–layer control of convection is never important. For example, isolated scattered deep convection over intense surface heating, such as over land, may be more directly influenced by the boundary layer processes. There is also an indication that the latter principle works rather well for shallow convection in practice [36]. Note that Bechtold *et al.* [10] also maintain a boundary–layer based closure for shallow convection.

T1.2: Critical Review of the Concept of Convective Quasi-Equilibrium

Convective quasi–equilibrium, as originally proposed by Arakawa and Schubert in 1974 [37], is considered one of the basic concepts in convection closure. A review by Yano and Plant [38], completed under the present Action, elucidates the richness of this concept with extensive potential possibilities for further investigating it from various perspectives. Especially, there are two contrasting possibilities for interpreting this concept: under a thermodynamic analogy, as originally suggested by Arakawa and Schubert, or as a type of slow manifold condition (or a balance condition [39]). The review suggests that the latter interpretation may be more constructive.

The review also suggests the importance of a more systematic observational verification of Arakawa and Schubert's hypothesis in the form that was originally proposed. Surprisingly, such basic diagnostic studies are not found in the literature in spite of their critical importance for more basic understanding of this concept. For example, though Davis *et al.* [40] examine convective quasi–equilibrium, their formulation is based on the re–interpretation of the Arakawa and Schubert's original formulation into a relaxation process (*cf.* Section 4.5 of [38]).

Furthermore, along a similar line of investigation, the convective energy cycle becomes another issue to be closely examined [9,41,42]. **Q1.2.1** discusses this issue further, but in short, Arakawa and Schubert's convective quasi–equilibrium is defined as a balanced state in the cloud–work function budget, which constitutes a part of the convective energy cycle. Importantly, the results from the energy–cycle investigations suggest that the concept of convective quasi–equilibrium could be more widely applicable than is usually supposed, with only a minor extension.

The principle of convective quasi–equilibrium is often harshly criticized from a phenomenological basis. For example, the importance of convective life–cycles is typically emphasized, which is not considered under quasi–equilibrium. The issue of self–organized criticality to be discussed in **Q1.7** could be a more serious issue.

However, ironically, we have never made a fully working quasi–equilibrium based convection parameterization. Consider a precipitation time series generated by an operational convection parameterization, ostensibly constructed under the quasi-equilibrium hypothesis: it is often highly noisy, suggesting that the system as operationally formulated does not stay as a slow process as the hypothesis intends to maintain (*cf.* Figure 6 [43]). In other words, operational quasi–equilibrium based convection parameterizations are not working in the way that they are designed to work. Making a quasi–equilibrium based parameterization actually work properly is clearly a more urgent issue before moving beyond the quasi–equilibrium framework. At the very least we need to understand why it does not work in the way intended.

Q1.2.1: How Can the Convective Quasi-Equilibrium Principle be Generalized to a System Subject to Time-Dependent Forcing? How Can a Memory Effect (e.g., from a Convection Event the Day before) Possibly be Incorporated into Quasi-Equilibrium Principle?

The importance of the issue raised here cannot be overemphasized. Even 40 years after the publication of the original article by Arakawa and Schubert in 1974 [37], it is very surprising to find that their original formulation is hardly tested systematically in the literature, as already mentioned in **T1.2** above. Though we are sure that there are lot of technical tests performed at an operational level, none of them is carefully reported in the literature.

This issue can be considered at two different levels. The first is a more direct verification of Arakawa and Schubert's convective quasi-equilibrium hypothesis (their Equation (150)) from observations. Here, the hypothesis is stated as

$$-\sum_{j=1}^{N}\mathcal{K}_{ij}M_{Bj} + F_i = 0 \tag{3}$$

for the i–th convective type, where $\mathcal{K}_{ij}M_{Bj}$ is a rate that the j–th convective type consumes the potential energy (or more precisely, cloud work function) for the i–th convective type, M_{Bj} is the cloud–base mass-flux for the j–th convective type, and F_i is the rate that large–scale processes produce the i–th convective–type potential energy. Here, N convection types are considered. The matrix elements, $\mathcal{K}_{ij}$, are expected to be positive, especially for deep convection due to its stabilization tendency associated with warming of the environment by environmental descent.

Currently intensive work by Jun-Ichi Yano and Robert S. Plant on this issue is underway. An important preliminary finding is that some of the matrix elements, $\mathcal{K}_{ij}$, can be negative due to a destabilization tendency of shallow convection associated with the re–evaporation of detrained cloudy air.

Second, the convective quasi-equilibrium principle can be generalized into a fully-prognostic formulation, as already indicated by Arakawa and Schubert [37] themselves, by coupling between an extension of their closure hypothesis (Equation (1.3), or their Equation (150)) into a prognostic version (*i.e.*, Equation (142))

$$\frac{dA_i}{dt} = -\sum_{j=1}^{N}\mathcal{K}_{ij}M_{Bj} + F_i \tag{4}$$

and the kinetic energy equation

$$\frac{dK_i}{dt} = A_i M_{Bi} - D_{K,i} \tag{5}$$

presented by their Equation (132). Here, A_i is the cloud work function, K_i the convective kinetic energy for the i–th convective type and the term $D_{K,i}$ represents the energy dissipation rate. Randall and Pan [44,45] proposed to take this pair of equations as the basis of a prognostic closure. The possibility is recently re-visited by Plant and Yano [9,41,42] under a slightly different adaptation.

This convective energy–cycle system, consisting of Equations (4a,b), describes evolution of an ensemble of convective systems, rather than individual convective elements. It can explain basic convective processes: e.g., convective life–cycles consisting of discharge (trigger) and recharge (suppression and recovery: [41]), as well as transformations from shallow to deep convection [9,42]. This result contains a strong implication, because against a common perception, the model demonstrates that an explicit trigger condition is not an indispensable ingredient in order to simulate a convective life cycle. Here, the life cycle of a convective ensemble is simulated solely by considering a modulation of convection ensemble under an energy-cycle description, keeping the evolution of individual convective elements implicit.

This energy-cycle formulation (4a,b) is, in principle, straightforward to implement into any mass-flux based convection parameterization, only by switching the existing closure without changing the entrainment-detrainment formulation. Most of the current closures take an analogous form to Equation (3), which may be generalized into a prognostic form (4). This is coupled with Equation (5), which computes the convective kinetic energy prognostically. The latter equation is further re–interpreted as a prognostic equation for the mass flux, by assuming a certain functional relationship between K_i and $M_{B,i}$. Here, a key is to couple shallow and deep convection in this manner, which are typically treated independently in current schemes. Many prognostic formulations for closure have already been proposed in the literature in various forms, e.g., [46]. However, it is important to emphasize that the formulation based on the convective energy cycle presented herein is the most natural extension of Arakawa and Schubert's convective quasi–equilibrium principle to a prognostic framework.

Q1.2.2: Are There Theoretical Formulation Available that could be Used to Directly Test Convective Quasi-Equilibrium (e.g., Based on Population Dynamics)?

Clearly this question is inspired by a work of [47], which suggests that it is possible to derive a population dynamics system starting from Arakawa and Schubert's spectrum mass-flux formulation (however see **T2.4**, [48]). Thus, it is also natural to ask the question other way round: can we construct and test a closure hypothesis (e.g., convective quasi-equilibrium) based on a more general theory (e.g., population dynamics)?

We have turned away from this direction during the Action for several reasons: (1) so far we have failed to identify a robust theoretical formulation that leads to a direct test of convective quasi-equilibrium or any other closure hypothesis; (2) it is dangerous to introduce an auxiliary theoretical condition to a parameterization problem without strong physical basis. This can make a parameterization more *ad hoc*, rather than making it more robust; (3) the convective quasi-equilibrium can better be tested in a more direct manner based on Arakawa and Schubert's original formulation as concluded in response to **Q1.2.1**.

T1.3: Proposal for a General Framework of Parameterization Closure

A closure condition is often derived as a stationarity condition of a vertically-integrated physical quantity. The two best known choices are water vapor (e.g., [49]) and the convective parcel buoyancy (e.g., [37]). The latter leads to a definition of CAPE or the cloud work function, more generally. (See **Q1.3.1** for an alternative possibility.)

A formulation for a closure under a generalization of this principle has been developed [50]. It is found that regardless of the specific choice of a physical quantity (or of any linear combinations of those), the closure condition takes the form of a balance between large-scale forcing, F_i, and convective response, $D_{c,i}$, as in the case of the original Arakawa and Schubert's quasi-equilibrium hypothesis, so that the closure condition can be written as

$$F_i + D_{c,i} = 0 \qquad (6)$$

for the i-th convective type. It is also found that the convective response term takes a form of an integral kernel, or a matrix, $\mathcal{K}_{i,j}$, in the discrete case, describing the interactions between different convective types, as in the Arakawa and Schubert's original formulation based on the cloud-work function budget. As a result, the convective response is given by

$$D_{c,i} = \sum_j \mathcal{K}_{i,j} M_{B,j} \qquad (7)$$

Arakawa and Schubert's convective quasi–equilibrium principle Equation (3) reduces to a special case of Equation (6). This general framework is expected to be useful in order to objectively identify basic principles for choosing more physically based closure conditions.

Note that though stochasticity may be added to a closure condition, it is possible only after defining a deterministic part of closure. See **Q1.3.2**, **T3.3** for further discussions (see also [51]).

Q1.3.1: Is It Feasible to Re-Formulate the Closure Problem as that of the Lower Boundary Condition of the System? Is It Desirable to Do So?

Formally speaking, the closure problem in mass-flux parameterization is that of defining the mass-flux at the convection base (*cf.* Section 2.1.1), and thus it may also be considered as a boundary condition. For example, the UM shallow-convection scheme is closed in this matter by taking a turbulent velocity measure as a constraint ([52]: see also [53,54]). However, as already discussed in Section 2.1.1, it is rather misleading to take the mass-flux closure problem as a type of bottom boundary condition, because what we really need is a general measure of convective strength, independent of any reference height, after a mass-flux vertical profile is normalized in a certain manner. It is just our "old" custom normalizing it by the convection-base value, but there is no strong reason to do so, especially if the mass-flux profile increases substantially from the convection base. This is the same reason as more generally why it is rather misleading to consider convection to be controlled by the boundary layer as already emphasized in **T1.1**. See [19] for more.

Q1.3.2: How does the Fundamentally Chaotic and Turbulent Nature of Atmospheric Flows Affect the Closure of Parameterizations? Can the Quasi-Equilibrium still be Applied for These Flows?

The convective energy cycle system, already discussed in **Q1.2.1**, is also the best approach for answering this question. In order to elucidate a chaotic behavior we have to take at least three convective modes. Studies have examined the one and two mode cases so far [9,42]. A finite departure from strict convective quasi-equilibrium may also be considered a stochastic process. Such a general framework, the method of homogenization, is outlined, for example, by Penland [55]. Specific examples of the applications include Melbourne and Stuart [56], and Gottwald and Melbourne [57].

T1.4: Review on Current State–of–the–Art of Entrainment-Detrainment Formulations

Entrainment and detrainment are technical terms referring respectively to the rate that mass enters into convection from the environment, and exits from convection to the environment (*cf.* Equation (2)). A review on these processes is published as [58] under the present Action.

T1.5: Critical Review of Existing Methods for Estimating Entrainment and Detrainment Rates from CRM and LES

As addressed in de Rooy *et al.* [58], there are two major approaches:

(1) Estimate by directly diagnosing the influx and outflux through the convection–environment interfaces [59,60]. Both of these studies use an artificial tracer for identifying the convection–environment interfaces.

(2) Less direct estimates based on a budget analysis of a thermodynamic variable [61,62]. The distinction between convection and environment is made by a threshold based criterion (vertical velocity, cloudiness, buoyancy, or a combination of those).

Unfortunately, these two approaches do not give the same estimates, but the former tends to give substantially larger values than the latter. The result suggests that we should not take the notion of the entrainment and the detrainment rates too literally, but they have meaning only under a context of a budget of a given variable that is diagnosed. Strictly, the latter estimate depends on a choice of a variable, as suggested by Yano *et al.* [63]. Also there is a subtle, but critical difference between the methods by [61,62], as discussed immediately below.

It should be emphasized that the second approach is based on an exact formulation for a budget of a given transport variable (temperature, moisture) derived from an original full LES–CRM system without any approximations. Thus, if a parameterization scheme could estimate all the terms given under this formulation, a self–consistent evaluation of convective vertical transport would be possible. However, the main problem is that it is hard to identify a closed formulation that can recover such a result. The current mass–flux formulation is definitely not designed in this manner. For the very last reason, neither approach gives entrainment and detrainment rates that lead to a mass flux profile that can predict vertical transport of a given variable exactly under a mass–flux parameterization. To some extent, Siebesma and Cuijpers [61] make this issue explicit by including a contribution of an eddy convective transport term (a deviation from a simple mass–flux based estimate) as a part of the estimation formula.

Swann [62], in turn, avoids this problem by taking an effective value for a convective component obtained from a detrainment term, rather than a simple conditionally–averaged convective value. As a result, under his procedure, the convective vertical flux is exact under the prescribed procedure for obtaining entrainment and detrainment in combination with the use of the effective convective value. Nevertheless, a contribution of environmental eddy flux must still be counted for separately. Furthermore, introduction of the effective value makes the convective–component budget equation inconsistent with the standard formulation, though a difference would be negligible so long as a rate of temporal change of fractional area for convection is also negligible.

Any of the estimation methods (whether direct or indirect) are also rather sensitive to thresholds applied for the distinction between convection and the environment. For example, Siebesma and Cuijpers [61] show that the values of entrainment and detrainment rates vary by 50% whether considering convection as a whole or only its core part (defined to have both positive vertical velocity and buoyancy). This sensitivity stems from the fact that when convection as a whole is considered, the differences between cloud and environment are smaller than if using the cloud core. Under this difference, in order to recover the same *total* convective vertical transport, $M\varphi_c$, for an arbitrary physical variable, φ, which is defined by a vertical integral of

$$\frac{\partial \varphi_c}{\partial z} = -\frac{E}{M}(\varphi_c - \bar{\varphi}) \tag{8}$$

we have to assume different mixing coefficients (fractional entrainment rate), E/M, depending on this difference, $\varphi_c - \bar{\varphi}$, between convection (core or cloud) and the environment. The same argument follows when a more direct estimate of entrainment–detrainment rates is performed. Note that in some convective schemes, the *eddy* convective transport, $M'(\varphi_c - \bar{\varphi})$, is considered in terms of the eddy convective mass flux, M', instead. Also note that being consistent with the analysis methods in concern, we assume only one type of convection in this discussion, dropping the subscript i for now.

Here, we should clearly distinguish between the issue of diagnosis based on LES–CRM and the computations of a convective profile within a parameterization. In the former case, all the terms are simply directly diagnosed (estimated) from LES–CRM output, and thus a self–consistent answer is obtained automatically. On the other hand, in running a parameterization, none of those terms are known *a priori*, thus they must somehow be all diagnosed (computed) in a self–contained manner without referring to any extra data. Clearly, the latter is much harder.

The most fundamental reason that these LES–CRM based entrainment–detrainment estimates do not find a unique formulation nor a unique choice of threshold is that those CRMs and LESs do not satisfy a SCA (segmentally–constant approximation) constraint that is assumed under the mass–flux formulation, as discussed later in **Q1.8**. In principle, better estimates of entrainment–detrainment would be possible by systematically exploiting a model under SCA but without entrainment–detrainment hypothesis. However, such a possibility is still to be fully investigated (*cf.* [64]).

An alternative perspective to this problem is to add an additional vertical eddy transport term estimated by a turbulence scheme. This perspective is consistent with the formulation proposed by Siebesma and Cuijpers [61], who explicitly retain the eddy transport in their diagnosis. This idea is further developed into convection parameterization combining the eddy diffusion and mass flux (EDMF: [54], see further **T4.3**). Of course, the turbulence scheme must be developed in such a manner that it can give an eddy–transport value consistent with an LES–CRM diagnosis. This is another issue to be resolved (*cf.* **T4.3**).

Q1.5.1: From a Critical Review of Existing Methods for Estimating Entrainment and Detrainment Rates from CRM and LES, What are the Advantages and Disadvantages of the Various Approaches?

The approach by Romps [59], and Dawe and Austin [60] presumably provides a more direct estimate of the air mass exchange rate crossing a convection–environment interface. However, there are subtle issues associated with the definition of the convection–environment interface, and how to keep track of it accurately.

First note that to an inviscid limit (*i.e.*, no molecular diffusion), which is a good approximation in convective scales, there would be strictly no exchange of air mass by crossing an interface defined in a Lagrangian sense. Such an interface would simply be continuously distorted with time by a typical turbulent tendency of stretching and folding into an increasingly complex shape, presumably leading to a fractal. Note that the turbulent processes only lead to continuous distortions of the material surfaces without mixing in the strict sense, though it may look like a mixing (as represented as an eddy diffusion) under a coarse graining. Such an interface evolution would numerically become increasingly intractable with time, with increasingly higher resolutions required. Clearly the computation results would be highly dependent on the model resolution.

An interface between convection (cloud) and the environment does not evolve strictly in a Lagrangian sense, but as soon as the cloud air evaporates, the given air is re–classified as an environment, and *vice versa*. Such a reclassification is numerically involved by itself, and the result would also sensitively depend on a precise microphysical evaluation for evaporation. In this very respect, we may emphasize the importance of returning to laboratory experiments in order to perform measurements not contaminated by numerical issues. Indeed a laboratory experiment can reveal many more details of entrainment–detrainment processes than a typical LES can achieve (Figure 1: *cf.* [65] see also [66]).

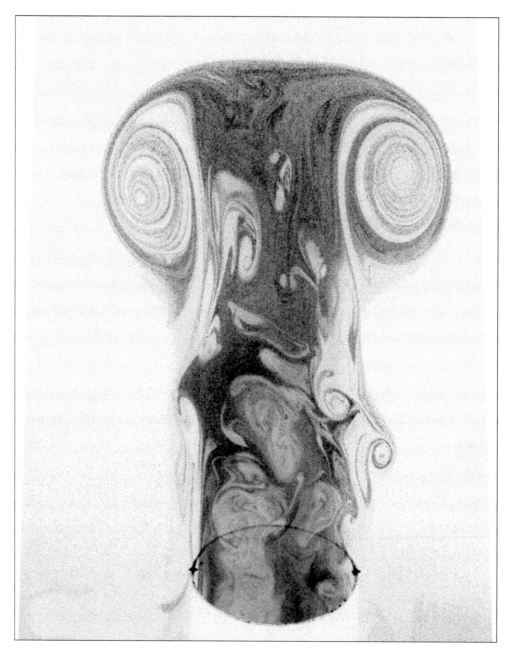

Figure 1. A cross section of a thermal plume generated in a laboratory with use of a humidifier as a buoyancy source. Distribution of condensed water is shown by gray tone (courtesy: Anna Gorska and Szymon Malinowski).

A cautionary note should be raised in conjunction with laboratory experiments in relation to the life–cycle issues discussed previously. An LES–CRM simulation will produce many cumulus clouds and an estimation of entrainment/detrainment rates across the full simulation effectively produces an average over the individual cloud life–cycles in a manner that is well suited to the consideration of an ensemble of clouds within a parameterization. The laboratory experiments focus on a single, isolated, buoyant plume, which is at once to their great advantage and disadvantage. See **Q1.6.1** for related issues.

See **T1.5** for further comparisons of the entrainment–detrainment evaluation methods.

T1.6: Proposal and Recommendation for the Entrainment-Detrainment Problem

The most important general suggestion drawn from [58] is an extensive use of CRM and LES in order to systematically evaluate the entrainment and the detrainment rates so that an extensive data base can be developed. As discussed in **T1.5**, such methodologies have already been well established.

Extensive LES studies for shallow convection have established that mixing between convection and the environment is dominated by lateral mixing across the convection–environment interfaces rather than a vertical mixing from the convection top, as proposed by Squires [67,68] and Paluch [69]. In this manner, it also establishes that the current entrainment–detrainment formulation for defining convective mass flux, M, under the Formula (2) is more robust than other existing proposals.

Furthermore, de-Rooy *et al.* [58] suggest some specific research directions:

(i) Critical fractional mixing ratio originally introduced in a context of a buoyancy sorting theory [70]: the critical fraction is defined as the mixing fraction between convective and environmental air that leads to neutral buoyancy. Mixing with less or more environmental air from this critical fraction leads to positive or negative buoyancy respectively. This division line is expected to play an important role in entrainment–detrainment processes.

(ii) Relative-humidity dependence of entrainment-detrainment rate: the introduction of such dependence clearly improves model behavior (e.g., [71,72]), although the mechanism behind is not yet well understood.

The vertical mass–flux profile is strongly controlled by detrainment as originally pointed out by [73]: see also [72,74]) with theoretical arguments provided by [75]. This finding has important consequences for the parameterization of convection: the critical mixing fraction correlates well with the detrainment rate, providing a possibility for taking it as a key parameter [73,74,76]. Unfortunately, almost all of the current parameterizations do not yet take this aspect into account. For example, the Kain and Fritsch scheme [70] assumes that entrainment and detrainment vary in opposite senses as functions of the critical mixing fraction. Some schemes have just begun to take this effect fully into account [73,76].

In spite of this progress, as a whole, unfortunately, we clearly fail to identify any solid theoretical guiding principle for investigating the entrainment–detrainment problem. The main problem stems from the fact that a plume is a clear oversimplification of atmospheric convection, as discussed in **Q1.6.1** next.

Q1.6.1: What is the Precise Physical Meaning of Entrainment and Detrainment?

The concept of entrainment is best established under the original context of the entraining plume experiment performed with a water tank by Morton *et al.* [77]. Once we try to extend this concept to moist atmospheric convection, we begin to face extensive controversies, some of which are discussed in de-Rooy-*et al.* [58]. In the moist convection context, even a precise physical meaning of the entrainment-detrainment concept is lost, as emphasized by Morton [78]. See Yano [79] for a historical review with extensive references therein. It simply reduces to a method for calculating lateral (and sometimes vertical) mixing crossing the boundaries of the air that is designated as convection.

The original entrainment-plume model is based on a premise that the convective plume has a relatively well-defined boundary against the environment, also approximately fixed with time. This idea is schematically well represented in Figure 2 of de Rooy and Siebesma [75]. This basic premise is also well summarized, for example, in the introduction of [80] by comparing this concept (as termed "entraining jet" in this paper) against the concept of a bubble or thermal. Here, in spite of the recent trend of more emphasizing the detrainment of air from the convective plume, this "jet" idea, in the sense of assuming a well-defined boundary with the environment, has hardly changed since then.

Another important premise, along with the existence of a well-defined boundary, is that the lateral exchange of the air between the convective plume and the environment is performed by eddies of scales much smaller than that of the plume itself. This idea is also schematically well represented by Figure 2 of [75]. The massive detrainment at the cloud top of individual clouds also contributes to the lateral exchange in an important manner.

However, do the true atmospheric moist convective systems actually behave in this manner? Our interpretation of Doppler radar measurements of winds given, for example, by Figure 3 of [81], may provide hints to this question. Here, keep in mind that Doppler radar is typically sensitive to precipitation, not cloud-particles, whenever there is precipitation present. Thus, one cannot infer cloud-edge sharpness with most Doppler radar in precipitating convection.

Probably the most striking feature of this convective element captured by a series of Doppler radar images with a frequency of every few minutes is its transient behavior without representing any fixed boundary in time. Furthermore, the flows around a convective cloud are subjectively rather "laminar". Within the limit of resolution of these radar measurements, we do not see any turbulent–looking "eddies" around the cloud. Instead, these "laminar" flows appear to provide extensive exchange of mass between the cloud and the environment. The frame (d') in their figure is probably the best one to make this point with a well-defined laminar inflow at a middle level.

Examination of these images does not exclude an obvious possibility that there are extensive turbulent eddies contributing to mixing at the scales unresolved by the radar. However, it is hard to believe that these unresolved small–scale eddies are responsible for most of the mixing between convection and the environment. This doubt is particularly justified by the fact that the whole cloud shape changes markedly over time without a well–defined fixed convection–environment interface.

A three-dimensional animation of a boundary-layer convective cloudy plume prepared by Harm Jonker and his collaborators [82] also makes the same point: highly transient nature of the convective plume, from which it appears hard to justify the traditional "steadiness" hypothesis of the convective plume. Based on such observation, Heus *et al.* [83] emphasize the existence of a "buffer zone" (descending shell) between the plume core and the environment. This "buffer zone" appears to roughly correspond to a "fuzzy" boundary of a cloud identified in terms of a high water–vapor concentration. The concept also appears to be consistent with the interpretation presented above that there is no boundary fixed in time between the cloud and the environment. Keep in mind that the boundary between the cloud (*i.e.*, visible cloud-particles) and environment is perfectly sharp and extremely well defined at any instant. The main issue here is that it fluctuates in time and has a fractal structure.

Of course, the argument above is slightly misleading in the sense that almost everyone would agree with a highly transient nature of realistic atmospheric convection. Moreover, many would also argue that the approximation of convection by a "steady plumes" adopted for the mass–flux convection parameterization, which also leads to a separation of the whole problem into closure and entrainment–detrainment (*cf.* Equation (1)), is a picture emerging only after an ensemble average of those individual clouds that has an equilibrium solution. In other words, schematics such as Figure 2 of [75] should not be taken too literally. Thus, the real question would be how to re–construct a steady plume solution under an ensemble averaging procedure for those transient convective clouds with their interfaces with the environment continuously changing with time. Such a systematic procedure is still needed.

Q1.6.2: If They Provide Nothing Other Than Artificial Tuning Parameters, How could they be Replaced with More Physically-Based Quantities?

Unfortunately, this question is not well posed. On the one hand, entrainment and detrainment are well–defined quantities that can be diagnosed objectively from CRM and LES, as already emphasized in **T1.5**. In this very respect, entrainment and detrainment are far from artificial tuning parameters, but clearly physically given. On the other hand, as emphasized in **Q1.6.1**, the basic physical mechanism driving these entrainment–detrainment processes is far from obvious. At least, the original idea of entrainment proposed by Morton *et al.* [77] for their laboratory convective plumes does not apply to atmospheric convection in any literal sense. Without such a theoretical basis, it may be rather easier to treat them purely as tuning parameters than anything physically based.

An approach that may be more constructive, however, can also be pursued under a variant of the conventional mass–flux framework. Note first that the convective profiles can be evaluated knowing only the entrainment rate, without knowing the detrainment rate, as seen by Equation (8). Second, recall that the mass flux, M, consists of two parts: convective vertical velocity, w_c, and the fractional area, σ_c, for convection,

$$M = \rho \sigma_c w_c \tag{9}$$

Thus, if we could compute these two quantities separately, there would no longer be a need for knowing entrainment and detrainment rates for use in computing the mass flux via Equation (2).

The convective vertical velocity is commonly evaluated by taking Equation (15) of [84]. Clearly this is a historical misquotation, because the formula in concern is derived for a spatially isolated spherical bubble, but not for a steady plume. Nevertheless, this use may be considered a necessary evil in order to evaluate the convective vertical velocity under a mass–flux formulation. The equation is formulated without entrainment and detrainment, at least in an explicit manner. (Levine's drag coefficient can be equated to the fractional entrainment rate if the same derivation is repeated under a more formal application of the mass–flux formulation (and SCA: *cf.* [18]).)

The fractional area for convection could, in turn, be evaluated by a fully–prognostic version of Equation (2):

$$\frac{\partial \sigma_c}{\partial t} = E - D - \frac{\partial M}{\partial z} \tag{10}$$

Of course, this equation retains both entrainment and detrainment. However, Gerard and Geleyn [85] were able to overcome this difficulty by introducing an alternative equation for σ_c based on the moist–static energy budget (their Equation (11)). In this manner, the mass flux can be evaluated without knowing entrainment and detrainment rates. Note that once the mass flux is known, the entrainment rate may be diagnosed backwards under certain assumptions. (Entrainment would still be required to compute the vertical profile of in-convection variables φ_c via Equation (8)).

A similar idea can be pursued with a further generalization of the mass flux framework into NAM-SCA, and effectively viewing the convection parameterization as a numerical issue rather than anything physical. In order to represent subgrid-scale convection, we do not want to have too strong convective vertical velocity (or too weak either). In order to control the degree of convective vertical velocity in a desirable manner (from a numerical point of view, in order to make the computations smooth), we need to adjust the fractional area for convection so that convection becomes neither too strong or too weak. Such an adjustment can be performed with a relatively simple numerical procedure without explicitly invoking an entrainment and detrainment rates. Such a formulation is relatively straightforward within NAM–SCA, as will be discussed in **Q1.8** below.

Q1.7: How Strong and How Robust is the Observational Evidence for Self-Organized Criticality of Atmospheric Convection?

Empirical studies across a broad range of observational scales have been attempted to characterize aspects of convective phenomena in order to constrain convective parameterizations, especially the closure. Critical properties are identified empirically, which may connect the convection parameterization problem with statistical physics theories of critical phenomena (*cf.* [86]). A broad range of atmospheric phenomena present scale-free distributions. Particularly, many atmospheric phenomena related to precipitation are associated with many characteristic temporal and spatial scales and present long-range correlations, which may result from the coupling between nonlinear mechanisms at different scales [87].

Peters *et al.* [88] analyzed high-temporal-resolution precipitation data and defined "episodic" precipitation events in a similar manner to avalanches in cellular-automaton models. It was found that a distribution of the precipitation event sizes (integrated rain rate over duration of the event) follows a power law over several orders of magnitude. A power-law distribution suggests criticality, but it is not a sufficient condition because trivial non–critical mechanisms can also lead to power laws [89].

Peters and Neelin [90] provided further evidence using TRMM satellite data over tropical oceans. A relationship between the satellite–estimated precipitation and the column–integrated water vapor is compatible with a continuous phase transition, in which large areas enter a convectively active phase above a critical value of column–integrated water vapor. Furthermore, they showed that precipitation events tend to be concentrated around the critical point. The precipitation variance was also found the largest around this point. These results can be interpreted in terms of a departure from quasi–equilibrium, and its scale-free behavior is consistent with the self-organized criticality (SOC). Furthermore, Peters-*et al.* [91] verified another expectation from the SOC framework, *i.e.*, a similarity of power-law exponents independent of the locations by using high temporal–resolution precipitation data. Data from the tropics was also found to exhibit an approximate power-law decay in auto–lag correlation [92]. A size distribution of mesoscale convective clusters also follows a power law [93,94]. These results suggest criticality of the atmospheric convective system, although alternative explanations for the observed behaviors are also possible. For example, a theory based on a stability threshold for boundary-layer water vapor is able to reproduce some aspects of the observed characteristics [95].

Peters and Neelin's results [90] appear to be robust, except for it is not clear how the relationship actually looks like above the critical point. The retrieved rain rates may be underestimated by TRMM microwave (TMI) due to a wrong retrieval method [96,97]. Further analysis is needed in order to confirm that the average precipitation is bounded for high water vapor values. Yano *et al.* [98] suggest, by analyzing an idealized planetary-scale convection simulation, that the shape of the relationship for upper values for precipitation would depend on a dependent variable chosen. For the column-integrated total water, as well as for condensed water, the results showed a similar tendency as in Peters and Neelin [90] but not for column-integrated water vapor. These two different tendencies were interpreted as indicative of two different underlying mechanisms: SOC and homeostasis.

Here, homeostasis is understood as a behavior of a system that keeps internal conditions rather stable in spite of external excitations. The system places a long delay before responding to an external excitation. It is almost indifferent to the excitation until a certain threshold is reached. Beyond that, however, the system responds with a high amplitude. A fast increase in the amplitude of reaction, just above a threshold can be considered as type of SOC. Under SOC, on the other hand, every sub-system of a given system has a threshold–dependent dynamics. Energy is accumulated (like grains on a column in the sand pile) and when a threshold is exceeded, a fast reaction (e.g., few grains are expelled) is induced in such manner that the sub-system returns to an equilibrium state, *i.e.*, a state under the threshold. There is a propagation of the effects to the nearest neighbors, which further associate sub-systems spatially connected together. This spatially extended events are called "avalanches". These avalanches extend over many different space scales, involving various sets of sub-systems. Such extensive involvements of subsystems lead to an allusion to "criticality". This is a major difference from the homeostasis, which only involves a single–scale system and is based on dynamical equations, sufficiently nonlinear to support threshold-type evolution. In invoking homeostasis to the physical picture of atmospheric convection, one must make sure that there is a kind of isolation of a sub-system so that it does not react to an external drive. Yano *et al.* [98] interpret Raymond's [99] thermodynamic self–regulation theory as a type of homeostasis (see also Section 6.3 of [38] for a review).

In practice, both SOC and criticality are the mathematical concepts that must be applied with great care to the real systems. Particularly, we should always keep in mind that theories are built upon for systems with infinite sizes, whereas the real systems have only finite sizes. For example, even for well–established cellular–automata SOC models, the relationship between tuning and order parameters can be substantially different from a standard picture discussed so far, as also found in second–order phase transitions in some cases [100].

As a whole, the evidence for atmospheric convective SOC still needs to be further investigated. This is challenging problem due to a lack of data for high water–vapor values. Future analysis, thanks to the advent of a new generation of satellite observation, such as the Global Precipitation Measurement (GPM) mission, may shade light on this issue. Along with the continuous observational investigations, large–domain LES/CRM simulations are also much encouraged.

Q1.8: Can a General Unified Formulation of Convection Parameterization be Constructed on the Basis of Mass Fluxes?

The mass-flux formulation can be considered to be built upon a geometrical constraint called segmentally constant approximation (SCA). This idea is first proposed by Yano-*et al.* [101], and further extended in [12,13,102,103]. Here, SCA is considered a basis for constructing a standard mass–flux formulation. For example, an application of SCA to a nonhydrostatic anelastic model is called NAM–SCAx [102].

A system purely constrained by SCA is general in the sense that any subgrid–scale processes that can be well represented under SCA would fit into this framework: such structures would include convective– and mesoscale updrafts and downdrafts, stratiform clouds, as well as various organized structures in the boundary layer such as cold pools.

A standard mass–flux parameterization can be derived from this prototype SCA model by adding three additional constraints:

(i) entrainment–detrainment hypothesis (*cf.* Section 2.1.1, **T1.5, Q1.5.1, T1.6, Q1.6.1, Q1.6.2**)

(ii) environment hypothesis: the hypothesis that all of the subgrid components (convection) are exclusively surrounded by a special component called the "environment"

(iii) asymptotic limit of vanishing fractional areas for convection, such that the "environment" occupies almost the whole grid–box domain.

The formulation structure of the mass–flux parameterization is carefully discussed in [12,13].

It may be important to emphasize that all these three constraints can be introduced without specifying whether subgrid–scale processes are convective or not, at least at a very formal level. The only real question is the degree to which a given subgrid–scale process can be described under these constraints. This also measures a degree of generality of mass–flux formulation.

At the same time, it is also emphasized that we can generalize the standard mass–flux formulation by removing some of the above constraints. In this manner, we can develop a general subgrid–scale parameterization by starting from the mass–flux formulation and then relaxing it in well-defined ways by removing or generalizing each of the standard constraints. From these perspectives, SCA provides a general framework for developing subgrid–scale parameterizations (*cf.* [12,13]).

Arakawa and Wu [104] outline a more general and universal framework for convection parameterization. Readers are strongly encouraged to critically examine these two strikingly different paths proposed here based on SCA and the one proposed by Arakawa and Wu (see also [105]).

2.2. *Non-Mass Flux Based Approaches: New Theoretical Ideas*

The key goal of this part of the Action activities is to identify useful new theoretical/mathematical ideas for convection parameterization development and studies. Specifically, we consider the approaches based on: Hamiltonian dynamics, similarity theories, probability density, and statistical mechanics. Unfortunately, MoU does not list stochasticity as an agenda. However, a book chapter [51] is exclusively devoted to this issue with extensive references therein (but see also **T1.3**, **Q1.3.2**, **T3.3**).

Q2.0: Does the Hamiltonian Framework Help to Develop a General Theory for Statistical Cumulus Dynamics?

The investigations of Hamiltonian dynamics and Lie algebra are major theoretical developments under the present COST Action [106–110]. In general, symmetries of differential equations are fundamental constraints on how physically self-consistent parameterizations must be constructed for a given system. Lie group analysis of systems of differential equations provides a very general framework for examining such geometrical properties of a system by means of studying its behavior under various symmetry transformations. The Hamiltonian framework furthermore simplifies these procedures. For the sake of structural consistency, the identified symmetries must also be preserved even when a parameterization is introduced to a system. This methodology can also be systematically applied to the mass-flux convection parameterization formulation so that fundamental theoretical constraints on the closure are obtained. This is considered an important future direction. See **Q2.3.5** for further.

T2.1: Review of Similarity Theories

Similarity theories, mostly developed in studies of turbulent flows, consist of two major steps: (1) perform a dimensional analysis so that a given system is nondimensionalized with a set of nondimensional parameters that characterize the behavior of the system; and then, (2) write down a nondimensional similarity solution that characterizes the system. In atmospheric science, this method is extensively exploited in boundary–layer studies generally for turbulent statistics, but more specifically for defining a vertical profile of vertical eddy fluxes. The latter are defined by a nondimensionalized profile function under the similarity theory. A book chapter is devoted to a review of this approach [111].

A particularly fascinating aspect of similarity theory is that, in principle, it contains the mass-flux formulation as a special case. Under this perspective, the mass flux formulation results from studying the Reynolds flux budgets, as shown by Grant [52,112] for shallow and deep convection, respectively. The similarity theory perspective furthermore suggests that approaches for convection parameterization are far from unique. In this very respect, this theory must be further pursued as an over-encompassing framework for all the subgrid-scale processes. Clearly this approach is currently under-investigated.

Note that the similarity theory is a particular choice for pursuing the first pathway identified under the present Action (*cf.* Section 1.3, [4]) by basing the convection parameterization development upon turbulence studies. Furthermore, the similarity theories may be considered a special case of nondimensional asymptotic expansion approaches. The latter perspective allows us to generalize similarity theories, which are primarily developed for steady states, to time–dependent problems. It can also be generalized from a point of view of moment expansions, which relate to Reynolds budgets.

Q2.1.1: What are the Key Non-Dimensional Parameters that Characterize the Microphysical Processes?

In fluid mechanics as well as in geophysical fluid dynamics, it is a standard procedure to nondimensionalize a system before making any investigations. The principal nondimensional parameters of a system are identified, and that in turn defines a phase space to explore in order to understand the behavior of the system. This approach is still far from a standard procedure for microphysical investigations. Possibilities for exploiting a dimensional analysis in a microphysical system in order to identify scaling relations are pursued by [113,114] for an idealized one–dimensional vertical model and an idealized orographical precipitation system, respectively. Here, their focus is in identifying the characteristic time scales of a given system. Stevens and Seifert [115] suggest how such characterizations may help to understand microphysical sensitivities in large–eddy simulations.

Yano and Phillips [116] provide a specific example for how a microphysical system can be nondimensionalized under an idealized zero-dimensional system, considering ice multiplication processes under ice-ice collisions. As it turns out, in this case, the whole behavior of the system is characterized by a single nondimensional parameter. The value of this nondimensional parameter can be estimated observationally, and thus the constructed phase diagram enables us to judge whether a given observed regime is under an explosive ice multiplication phase (a particular possibility identified in this study) or not. So far, only preliminary investigations have been performed. A full–scale investigation of the microphysical system under a systematic nondimensionalization is a promising direction, but one that is still to be taken.

Q2.1.2: How can the Correlation be Determined between the Microphysical (e.g., Precipitation Rate) and Dynamical Variables (e.g., Plume Vertical Velocity)?

In the four years of the present Action, we have declined to pursue this possibility. A correlation analysis is well known to be susceptible of producing misleading conclusions and it appears to us that it is difficult to construct a clean correlation analysis of the issue that would identify a useful physical causality. The most formal and robust manner for coupling between mircophysical processes and convective dynamics within a parameterization context is to define the convective vertical velocity consistently. The key issue from a microphysical point of view is that it is imperative to specify a vertical velocity distribution for a sub-convective scale in order to describe the microphysical processes properly (in a satisfying manner, as done within a CRM: *cf.* [17]). On the other hand, a standard bulk convection parameterization can only provide a single convective vertical velocity. The argument can easily move ahead to propose a crucial need for adopting a spectral description of parameterized convection.

However, this argument is likely to be rather short-circuited. First of all, a spectrum of convective plumes types does not provide a distribution for the sub-convective scales as required for proper microphysical descriptions. Second, the microphysics expects a time-evolving convective dynamics, whereas a standard mass-flux formulation only provides a steady solution by assuming a steady plume. Technically, a prognostic description of convection under the mass flux formulation is straightforward under its SCA extension (*cf.* **Q1.8**).

Lastly, and most importantly, what microphysicists would like to implement in a convection parameterization is rather an explicit microphysics, although it may well be phenomenologically developed (*cf.* Section 1.1). Efforts are clearly required to develop microphysical descriptions to a parameterized level so that, possibly, fine details of the convective dynamics may be no longer necessary (*cf.* **Q4.3.1**). The last point further leads us to a more general question: to what extent are microphysical details required for a given situation and a given purpose? Here, the microphysicists tend to emphasize strong local sensitivities to microphysical choices. On the other hand, the dynamicists tend to emphasize a final mean output. Such inclinations can point towards opposite conclusions for obvious reasons, and doubtless we need to find an appropriate intermediate position (*cf.* **Q3.4.3**).

Q2.1.3: How should a Fully Consistent Energy Budget be Formulated in the Presence of Precipitation Processes?

As we already emphasized in Section 1.3, more intensive investigations of the parameterization problem form the turbulence point of view are required. However, **Q2.1.3** is typical of the issues that must be addressed when this pathway is pursued. Purely from a point of view of mechanics, this is rather a trivial question: one performs a formal energy integral for the vertical momentum equation. Although it is limited to a linear case, the clearest elucidation of this method is offered by [117]. A precipitation effect would simply be found as a water-loading effect in the buoyancy term. This contribution would be consistently carried over to a final energy-integral result. Furthermore, the water–loading effect can be reintegrated to the "classical buoyancy terms" under a consistent formulation [16,118].

The real issue arises when this energy budget is re-written in the context of the moment expansion framework, on which similarity theory (*cf.* **T2.1**) is based. The moment-based subgrid-scale description has been extensively developed in turbulence studies, with extensive applications in the dry turbulent boundary layer. This theoretical framework works well when the whole process is conservative. Constructing such a strictly conservative theory becomes difficult for the moist atmosphere, due to the existence of differential water flux [16,118,119], and once a precipitation process starts, the whole framework, unfortunately, breaks down even under standard approximations.

On the other hand, invocation of the Liouville principle (*cf.* **T2.2** and the following questions) provides a more straightforward description for the evolution of the water distribution under precipitation processes so long as the processes are described purely in terms of a single macrophysical point. Note that the precipitation process itself would be more conveniently treated under a traditional moment-based description as a part of the eddy transport.

Hence, the turbulent–kinetic energy evolution under precipitation is best described under a time splitting approach: compute the traditional turbulent process (including water fall out) using a moment-based approach, and then update the microphysical tendencies based on the Liouville principle.

T2.2: Review of Probability-Density Based Approaches

See a book chapter [120], which reviews probability–density based cloud schemes. Clouds are highly inhomogeneous for a wide range of scales, and most of them are not well resolved in numerical models. Thus, a need arises for describing subgrid–scale cloud distributions. In the following, we are also going to focus on the issues of cloud schemes.

Q2.2.1: How can Current Probability-Density Based Approaches be Generalized?

The best general approach would be to take that of time splitting between the physics part and the transport part. The physics part (*i.e.*, single-point processes) is handled by the Liouville equation, as further emphasized in **Q2.2.5**. On the other hand, the transport part (eddy transport) is handled based on the moment-based description, invoking an assumed pdf approach (*cf.* **Q2.2.3**).

Q2.2.2: How can Convective Processes be Incorporated into Probability-Based Cloud Parameterizations? Can Suitable Extensions of the Approach be Made Consistently?

This question can be interpreted in two different ways: (1) a possibility of treating convection (or more precisely deep convective towers) as a part of a probability–based cloud scheme; or, (2) incorporate the effect of convection (especially deep convection) or interaction with deep convection as a part of a probability–based cloud scheme. In the latter case, convection is treated by a different scheme, say, based on mass flux, and it is *not* counted as a part of the cloud considered by the given probability–based scheme. In other words, the cloud scheme only deals with the so–called stratiform clouds. In pursuing the first possibility, the tail of the probability distribution becomes important, because deep convective towers tend to produce high water mixing ratios. In order to well account for the tail part of a distribution, higher–order moments must be included in a formulation. The inclusion of a skewness would be a minimum in order to take this step, and has been followed by, e.g., [121–123].

In particular, Bony and Emanuel [121] claim that the shape of the probability distribution is altered so that at every time step, the in–cloud value of cloud water equals the sum of those diagnosed by a traditional large–scale saturation and a convection parameterization. However, a careful examination of their formulation suggests that this statement is rather an understanding than anything actually derived as a formulation (*cf.* **T2.4**). In general, it is not obvious how to describe convective evolution in terms of higher–order moments (e.g., skewness) regardless of whether the issues are handled in a self–contained manner or under a coupling with an independent convection scheme. The difficulty stems from a simple fact that a spatially–localized high water concentration associated with deep convective towers is not easily translated into a quantitative value of skewness.

The second possibility is, in principle, more straightforward: the cloudy air detrained from convection is counted as an additional source term in a cloud scheme budget. This additional source term would be relatively easily added under a formulation based on the Liouville equation: the convective source enters as a flux term that shifts the water distribution from lower to higher values by extending the tail of the distribution. UM PC2 [124] takes into account the feeding of clouds from convection to stratiform, at least conceptually, in this manner, but without explicitly invoking the Liouville principle. Alternatively, Klein *et al.* [125] try to deal with this problem by considering moments (variances) associated with deep convection. However, this alternative approach is not quite practical for the mass–flux based convection parameterization, which does not deal with these variants by default. The convection parameterization would have to be further be elaborated for this purpose. The above Liouville–based approach, on the other hand, can handle the problem without explicitly invoking higher moments for convection.

Q2.2.3: Is the Moment Expansion a Good Approximation for Determining the Time-Evolution of the Probability Density? What is the Limit of This Approach?

This question specifically refers to an idea of assumed pdf originally developed by Golaz *et al.* [126]. Although this approach is attractive with a possibility of truncating the moments by post–analysis by CRM or LES for a given case, it appears hard to generalize it easily without further testing. Unfortunately, we fail to identify any suitable mathematical theorem for measuring the convergence of the pdf under a moment expansion.

Q2.2.4: Could the Fokker-Planck Equation Provide a Useful General Framework?

The Fokker–Planck equation is a generalization of the Liouville equation that is appropriate for certain stochastic systems. In other words, the Fokker–Planck equation reduces to the Liouville equation when the system is deterministic.

In the context of cloud parameterizations, we should note that the concept of "pdf" (probability density function) is slightly misleading, but it is better called "ddf" (distribution density function), because here we are dealing with a distribution of a variable (e.g., total–water mixing ratio) over a grid box rather than any probability (e.g., chance to find a condensed water at a given point). Current approaches for cloud schemes are, in principle, deterministic, although stochasticity may sometimes be added (with a possible confusion arising from the subtle distinction between pdf and ddf).

For issues of stochasticity itself, see: **T1.3,Q1.3.2, T3.3**, and Plant and Bengstsson [51].

Q2.2.5: How can Microphysics be Included Properly into the Probability-Density Description?

The Liouville equation is the answer, because it describes any single physical–point processes well, as already discussed in **Q2.2.1**. Here, the assumed pdf approach becomes rather awkward, because it is hard to include microphysical processes (a process conditioned by a physical–space point) into moment equations.

T2.3: Assessments of Possibilities for Statistical Cumulus Dynamics

Here, the statistical cumulus dynamics refers to the description of the cumulus dynamics in analogy with the statistical mechanics (*cf.* **Q1.7**). It is often argued that subgrid-scale parameterization is fundamentally "statistical" in nature (*cf.* [49]). However, little is known of the statistical dynamics for atmospheric convective ensembles. This is a domain that is clearly under–investigated. If we really wish to establish convection parameterization under a solid basis, this is definitely where much further work is needed. The present Action has initiated some preliminary investigations. Especially, we have identified renormalization group theory (RNG) as a potentially solid starting point [11]. We strongly emphasize the importance of more extensive efforts towards this direction.

Q2.3.1: How can a Standard, "Non-Interacting", Statistical Description of Plumes be Generalized to Account for Plume Interactions?

When this question was originally formulated, we did not fully appreciate the fact that the conventional spectrum mass-flux formulation *does* consider the interactions between convective plumes, albeit in an indirect sense, through the environment. Work with the Action has begun the analysis of the interactions between convection types within this framework, with a view to providing an assessment of whether or not such interactions have important implications and consequences for convection parameterization performance.

Perhaps the best example for making this point is the transformation from shallow to deep convection as elucidated for the two–mode mass–flux formulation (*cf.* [9,42]). Without mutual interactions, shallow convection is a self-destabilizing process associated with its tendency for moistening and cooling, whereas deep convection is a self-stabilizing process associated with its tendency for drying and warming. A proper coupling between these two types of convection is a key for properly simulating the transformation process. As already emphasized in **Q1.2.1**, this is a formulation that can be relatively easily implemented into operational models as well. Considering more direct interactions between convective elements is straightforward under the SCA framework (*cf.* **Q1.8**). A key missing step is to develop a proper statistical theory under this framework.

Q2.3.2: How can Plume-Plume Interactions and Their Role in Convection Organization be Determined?

Several steps must still be taken in order to fully investigate this issue under a framework of statistical mechanics. First is an extensive elementary study under the SCA framework. Second is a need for developing a proper Hamiltonian framework for the SCA system so that this system can be more easily cast into a framework suitable for statistical mechanics analyses under a Hamiltonian formulation (*cf.* **Q2.3.5**).

Q2.3.3: How can the Transient, Life-Cycle Behavior of Plumes be Taken Into Account for the Statistical Plume Dynamics?

Extensive statistics can be developed by examining both CRM and LES outputs of convection simulations [127,128]. The next question is how to develop a self-contained self-consistent statistical theory based on these numerically accumulated statistics. However, see **T1.1**, **Q1.2.1** for reservations for advancing towards this direction.

Q2.3.4: How can a Statistical Description be Formulated for the Two-Way Feedbacks between Convective Elements and Their "Large-Scale" Environment?

The convective energy-cycle description already discussed in **Q1.2.1** would be the best candidate for this goal. Technically, it is straightforward to couple this convective energy-cycle system with simple models for large-scale tropical dynamics. The is an important next step to take.

Q2.3.5: How can Statistical Plume Dynamics Best be Described Within a Hamiltonian Framework?

As is well known in statistical physics, once a Hamiltonian of a given system is known, various statistical quantities associated with this system can be evaluated in a straightforward manner through a partition function. There is no technical difficulty for developing such a Hamiltonian formulation (so long as we take nondissipative limit to a system) for an atmospheric convective system. Much extensive investments and funding are clearly required towards this goal.

T2.4: Proposal for a Consistent Subgrid-Scale Convection Formulation

In common scientific discourses, consistency of a given theoretical formulation presents two major distinct meanings:

(1) Self-consistency
(2) Consistency with physics

The first definition refers to the self-consistency of the logic when a formulation is developed in deductive systematic manner. We suggest to take the first definition for consistency for parameterization in order to avoid possible confusions discussed below.

An example of inconsistency in logic is, for example, found in Wagner and Graf [47], as pointed out by Plant and Yano [48]: it assumes both the cloud work function, A, and the mass flux, M, change in time with the same rate in the order

$$\frac{1}{A}\frac{\partial A}{\partial t} \sim \frac{1}{M}\frac{\partial M}{\partial t} \tag{11}$$

at a one point, and then the rate of change of the cloud work function is much slower than that of the mass flux, *i.e.*,

$$|\frac{1}{A}\frac{\partial A}{\partial t}| \ll |\frac{1}{M}\frac{\partial M}{\partial t}| \tag{12}$$

at another point under a single derivation process. The two conditions, Equations (8) and (12), are clearly contradicting each other. Thus the derivation is clearly not consistent. However, here and elsewhere, a value of a heuristic derivation should not be disputed. A consequence of a logical inconsistency is often hard to measure, and a practical benefit wins over.

On the other hand, some people take the word differently. In this second definition, the question is posed whether a given formulation is consistent with a given physics, or known physics. For example, the conventional mass–flux formulation for convection parameterization can be regarded as inconsistent because it does not take into account the role of gravity waves in convective dynamics. By the same token, quasi-geostrophic dynamics can also be regarded as inconsistent because it also does not take into account the contributions of gravity waves to the dynamics. It is debatable whether the first example is problematic, but in the second example, quasi–geostrophic dynamics is widely accepted despite this point.

In the second definition, we have to carefully define the relevant physics. Clearly, all the physical descriptions in atmospheric models do not take account of quantum effects, which are considered negligible for all the modeled processes. This type of inconsistency is not an issue. The role of gravity waves is more subtle, although it is still likely that in many situations they can be neglected.

From this point of view, this type of consistency is better re-interpreted in terms of the accuracy of an approximation adopted.

From a practical point of view, consistency of the thermodynamic treatment warrants special attention. Traditionally, atmospheric thermodynamics are often considered under various arbitrary approximations, and it is even difficult to examine the self–consistency in retrospect. One of our major achievements is to show how atmospheric thermodynamics can be constructed in a self–consistent manner [14–16,118,129]: see Section 2.3.1 for more. The relationships between the cloud and the convective schemes, already discussed in **Q2.2.2** provide a good example for further considering the issues of self–consistency of parameterization.

Two approaches were discussed in **Q2.2.2**. The first is to establish mutual consistency between the convective and the cloud schemes. "Consistency" here means a logical consistency by writing the same physical processes in two different ways within two different parameterizations. More precisely, in this case, an "equivalence" of the logic must be established. A classic example of such an equivalence of logic is found in quantum mechanics between the matrix–based formulation of Heisenberg and the wave–equation based formulation of Schrödinger. The equivalence of the formulations may be established by a mathematical transformation between the two. Such a robust equivalence is hardly established in parameterization literature.

The second approach is to carefully divide clouds into convective and non–convective parts, and let the convection and the cloud parameterizations deal each part separately. In this second case, an issue of double counting must be avoided. Here, a notion of dichotomy between convection and environment introduced by standard convection parameterizations becomes important. In order to avoid any double counting, the cloud scheme should deal with the clouds only in the environmental part and not in convective part. This is the basic principle of retaining mutual consistency of two physical processes: separate them into different subdomains over a grid box. The concept of SCA helps to handle this issue in lucid manner (*cf.* **Q1.8**).

A similar issue also arises in dealing with all types of subgrid–scale motions consistently. As suggested in the beginning of Section 2.1, our tradition is to deal with them under a dichotomy between convection and turbulence. However, in this case, too, a careful separation between them must be performed in order to avoid double counting. To some extent, it could be easier to deal with all types of motions under a single formulational framework (*cf.* **Q3.4.2** for more).

A corollary to this discussion is that, regardless of the decisions made about how consistency is to be achieved in a model, it is essential that the decision be made clearly, cleanly and openly. The developers of the individual parameterizations must all be agreed on the strategy. Moreover, model users should be aware that parameterization schemes are not necessarily interchangeable: a particular convection paramaterization should not be expected to function well if coupled to cloud or boundary layer schemes that do not share its assumptions about which scheme is treating which processes. The literature suggests that such awareness is not always as strong as we would wish.

2.3. High-Resolution Limit

As resolution increases both for weather–forecast and climate models, a number of new aspects must be addressed, especially the adequacy of the present convection parameterizations. Convection parameterization is traditionally constructed by assuming a smallness of the convective scales compared to a resolved scale (*i.e.*, scale separation principle). A parameterization scheme must somehow be adjusted based on the model resolution (*i.e.*, resolution–dependency). These are the issues to be addressed in the present section.

T3.1: Review of State-of-the-Art of High-Resolution Model Parameterization

See a book chapter [130] for a review, and **T.3.3** for further discussions.

T3.2: Analysis Based on Asymptotic Expansion Approach

We may consider that traditional parameterizations are constructed under an asymptotic limit of scale separation (the constraint iii) in **Q1.8**). For a parameter for the asymptotic expansion, we may take the fractional area, σ_c, occupied by convection. This is a standard small parameter adopted in mass-flux convection parameterization, which is taken to be asymptotically small.

However, as model resolution increases, this asymptotic limit becomes less valid. In order to address this issue, the exercise proposed here in the MoU was to move to a higher order in the expansion so that a more accurate description may be obtained. As an example, a higher–order correction to a standard mass-flux convection parameterization formulation was attempted. As it turns out, the obtained higher-order correction is nothing other than a particular type of numerical time-stepping scheme that makes the scheme weakly prognostic (see Appendix for the details). This is essentially consistent with the result obtained by more directly removing the asymptotic limit: the mass–flux formulation becomes fully prognostic as a result.

T3.3: Proposal and Recommendation for High-Resolution Model Parameterization

So far as the mass-flux based parameterization is concerned, a standard asymptotic limit of vanishing fractional convective area must be removed when a model resolution is taken high enough so that a standard scale separation is no longer satisfied. Thus, the formal answer to this issue is to make the mass-flux parameterization fully prognostic, also by taking out a standard "steady plume" hypothesis, as remarked in **T3.2**.

However, as far as we are aware, this fully drastic measure is not yet taken at any operational research centers so far. Several different approaches are under consideration.

The first approach is to stick to the standard mass–flux convection parameterization formulation based on an asymptotic limit of $\sigma_c \to 0$. This strategy, currently adopted at ECMWF [131], may be justified at the most fundamental level, by the fact that a good asymptotic expansion often works extremely well even when an expansion parameter is re-set to unity. Such a behavior can also be well anticipated for mass-flux convection parameterization. At the practical level, what the model can actually resolve (*i.e.*, the effective resolution) is typically more than few times larger than a formal model resolution, as defined by a grid-box size, due to the fact that a spatial gradient must be evaluated numerically by taking over several grid points.

Under this approach, a key missing element is lateral communication of convective variability between the grid boxes. As a partial effort for compensation of this defect, the convection parameterization has been coupled with a stochastic cellular automaton scheme. The latter mimics lateral interactions associated with convective processes in a very crude, but helpful manner [132,133].

The second approach is to move towards a more prognostic framework in an incremental manner under a framework of traditional parameterization. This effort is called 3MT (Module Multiscale Microphysics and Transport) [85,134,135]. Although it may be considered somehow "backwards" in a sense as going to be criticized in **T4.3** below, careful efforts are made in these studies to avoid any double counting in the interactions between the otherwise–competing computations of thermodynamic adjustment and convective latent heat release, as well as latent heat storage for downdrafts.

The third approach is to add a stochastic aspect to a standard scheme based on the quasi-equilibrium hypothesis in order to represent finite departures from quasi-equilibrium that can be expected in the high resolution limit [136]. An important technical detail under this implementation is the need for defining an effective environment larger than the grid–box size, over which the standard quasi-equilibrium assumption can reasonably be applied. The approach can therefore be considered as a downscaling of the convective response, which implies a stochastic formulation (*cf.* Section 2.4.2).

Efforts ongoing at DWD (TKE-Scalar Variance mixing scheme: TKESV, by Machulskaya and Mironov [137]) identify a key issue in the high-resolution limit as being an improvement of a boundary-layer scheme associated with cloud processes. Here, a major challenge is the inclusion of a proper water cycle in the context of a traditional turbulence parameterization (*cf.* **Q2.1.3**). A hybrid approach combining the traditional moment-based approach and a subgrid-scale distribution is adopted for this purpose. When a relatively simple distribution is pre-assumed for a latter, a closed formulation was developed relatively easily at DWD.

In reviewing these different approaches, it may be remarkable to note that the two approaches, 3MT and TKESV, adopt mutually consistent thermodynamic descriptions. On the other hand, these two efforts take contrasting approaches in dealing with the dichotomy between convection and turbulence. 3MT takes into account a gradual shift from convective to turbulent regimes with the latter being delegated in a self-consistent manner to the other parameterization schemes and to the dynamics. On the other hand, TKESV reduces the issues of all of the subgrid–scale motions to that of a turbulence problem (*cf.* **T2.4**).

Some further perspectives can be found in e.g., [104,105]).

2.3.1. More General and Flexible Parameterization at Higher Resolutions

The issue for making a parameterization general and flexible is best discussed under a universal setting. Issues at higher resolution would simply be considered a special case of this *general problem.* We even argue that this is a moral for modeling rather than any specific scientific issue. Thus, our following answer is also presented in such manner.

We propose the three basic dictums:

(1) Start from the basic laws of physics (and chemistry: *cf.* **T4.3**)

(2) Perform a systematic and logically consistent deduction from the above (*cf.* **T2.4**)

(3) Sometimes it may be necessary to introduce certain approximations and hypotheses. These must be listed carefully so that you would know later where you introduced them and why.

Atmospheric sciences are considered applications of the basic laws of physics (and chemistry). Since the Norwegian school established modern meteorology, it remains the basic principle of our discipline, because otherwise we lose a robustness in our scientific endeavor. Of course, not all the laws of physics are precisely known for atmospheric processes. Cloud microphysics would be the best example that must tackle with numerous unknowns. However, we must start from robust physics that we can rely upon. Another way to restate the first dictum above is: "never invent an equation". Thus, any development must start from robust known physics, and the uncertainty of our physical understanding of a given process must properly be accounted for in the development process of the parameterization (*cf.* **T4.3**).

A parameterization is, by definition, a parametric representation of the full physics that describes the subgrid scales. Thus, a certain deduction from the full physics is required in order to arrive such a parametric representation. Such a deduction process must be self–consistent and logical: a simple moral dictum. While simply said, in practice a completely self–consistent logical deduction is almost always not possible for many complex problems in parameterization. Certain approximations and hypotheses must inevitably be introduced. At a more practical level, thus those approximations and hypotheses must carefully be listed during the deduction process with careful notes about extent of their validity and limits. In this manner, we would be able to say how much generality and consistency is lost in the deduction process. Here, the main moral lesson is: be honest with these. Specific examples for developing a subgrid–scale parameterization in a general manner under the above strategy are: mode decomposition [101], moment expansion [138], and similarity theory ([111], *cf.* **T2.1**). SCA introduced in **Q1.8** may be considered a special application of mode decomposition.

The main wisdom stated above may be rephrased as "start from robust physics that we can trust and never reinvent a wheel". An unsuspected issue concerns the moist-air entropy, because although the liquid-water and equivalent potential temperatures are commonly used to compute the specific values and the changes in moist-air entropy, this is only valid for the special case of closed systems where total water content (q_t) and thus dry-air content ($q_d = 1 - q_t$) are constant for a moving parcel in the absence of sources and sinks. Marquet [14] proposes a more general definition of specific moist-air entropy which can be computed directly from the local, basic properties of the fluid and which is valid for the general case of barycentric motions of open fluid parcels, where both q_t and q_d vary in space and in time. Computations are made by applying the third law of thermodynamics, because it is needed to determine absolute values of dry-air and water-vapor entropies independently of each other. The result is that moist-air entropy can be written as $s = s_{ref} + c_{pd} \ln(\theta_s)$, where s_{ref} and c_{pd} are two constants. Therefore, θ_s is a general measure of moist-air entropy. The important application shown by Marquet and Geleyn [14–16,118,129] is that values and changes in θ_s are significantly different from those of θ_l and θ_e if q_t and q_d are not constant. This is especially observed in the upper part of marine stratocumulus and, more generally, at the boundaries of clouds.

Barycentric and open-system considerations show that the moist-air entropy defined in terms of θ_s is at the same time: (1) a Lagrangian tracer; and, (2) a state function of an atmospheric parcel. All previous proposals in this direction fulfilled only one of the two above properties. Furthermore, observational evidence for cases of entropy balance in marine stratocumulus shows a strong homogeneity of θ_s, not only in the vertical, but also horizontally: *i.e.*, between cloudy areas and clear air patches. It is expected that these two properties could also be valid for shallow convection, with asymptotic turbulent and mass-flux-type tendencies being in competition with diabatic heating rates.

Several implications are drawn here. First, the moist-air entropy potential temperature θ plays the double role of: (i) a natural marker of isentropic processes; and, (ii) an indirect buoyancy-marker (unlike in the fully dry case where the dry-air value θ directly plays such a role). Indeed, the Brunt-Väisälä frequency can be separated in terms of vertical gradients of moist-air entropy and total water content [16,118], almost independently whether condensation/evaporation takes place or not within parcels (simply because moist-air entropy is conserved for adiabatic and closed processes).

The main impact of moisture on moist-air entropy is the water-vapor content, which is already contained in unsaturated regions and outside clouds. The condensed water observed in saturated regions and clouds leads to smaller correction terms. An interesting feature suggested by Marquet and Geleyn [118] is that the large impact of water vapor does not modify so much the formulation of the Brunt-Väisälä frequency when going from the fully dry-air (no water vapor) to the "moist-air" (cloudy) formulations. These results indicate that parameterization schemes relying on phenomenological representations of the links between condensation/evaporation and microphysics might not be the only answer to the challenges discussed here.

From the entropy budget point of view (and hence perhaps also for the energy or enthalpy budget) the issue that matters is the presence of precipitation and entrainment/detrainment processes as generators of irreversibility and as witness of the open character of atmospheric parcels' trajectories. This is especially true for marine stratocumulus where clouds have comparable entropy to unsaturated patches and subsiding dry-air above [14,16]. Hence, a moist–turbulent parameterization schemes with a medium–level of sophistication must be based on the moist-air entropy. It is also important for such a scheme to have a reasonable and independent closure for the cloud amount, which ought to be competitive with the particular roles in organized plumes through eddy-diffusivity mass-flux schemes.

Note that two of the approaches for the high–resolution limit discussed in **T3.3** (3MT and TKESV) are perfectly compatible with this new type of thinking.

Q3.4: High–Resolution Limit: Questions

The following questions are listed for the high–resolution limit in MoU.

Q3.4.1: Which Scales of Motion should be Parameterized and under Which Circumstances?

A very naive approach to this problem is to examine how much variability is lost by averaging numerically–generated output data from a very high resolution simulation with a CRM or LES that well resolves the fine-scale processes of possible interest. The analysis is then repeated as a function of the averaging scale. In fact, this exercise could even be performed analytically, if a power–law spectrum is assumed for a given variable.

However, one should realize that whether a process needs to be parameterized or not cannot be simply judged by whether the process is active or above a given spatial scale. The problem is much more involved for several reasons:

(i) Any process in question cannot be characterized by a single scale (or wavenumber), but is more likely to consist of a continuous spectrum. In general, a method for extracting a particular process of concern is not trivial.

(ii) Whether a process is well resolved or not cannot be simply decided by a given grid size. In order for a spatial scale to be adequately resolved, usually several grid points are required. As a corollary of this, and of point (i), the grid size required depends on both the type of process under consideration and the numerics.

(iii) Thus the question of whether a process is resolved or not is not a simple dichotomic question.

With these considerations, it would rather be fair to conclude that the question here itself is ill posed. It further suggests the importance of a gradual transition from a fully parameterized to a well–resolved regime (*cf.* **T3.3**).

A way to override this issue could be to handle the issue of subgrid–scale parameterization like that of an adaptive mesh–refinement. A certain numerical criterion (e.g., local variance) may be posed as a criterion for mesh refinement. A similar criterion may be developed for subgrid–scale parameterizations. A conceptual link between the parameterization problem and numerical mesh–refinement is suggested by Yano *et al.* [102]. For general possibilities for dealing a parameterization problem as a numerical issue, see **Q1.6.2**.

Q3.4.2: How can Convection Parameterization be Made Resolution-Independent in order to Avoid Double-Counting of Energy-Containing Scales of Motion or Loss of Particular Scales?

There are two key aspects to be kept in mind in answering this question. First is the fact that basic formulations for many subgrid–scale parameterizations, including mass flux as well as an assumed pdf, are given in a resolution independent manner, at least at the outset. This is often a consequence of the assumed scale separation. It is either various *a posteriori* technical assumptions that introduce a scale dependence, or else the fact that the scale separation may hold only in a approximate sense. In the mass–flux formulation, major sources of scale dependence are found both in closure and entrainment–detrainment assumptions. For example, in the closure calculations a common practice is to introduce a scale–dependent relaxation time–scale, and satisfactory results at different model resolutions can only be obtained by adjusting such a parameter with the grid size. A similar issue is identified in Tompkins' [122] cloud scheme, which contains three rather arbitrary relaxation time–scales.

However, once these arbitrary relaxation time–scales are identified in a scheme, a procedure for adjusting them may be rather straightforward. From a dimensional analysis, and also particularly invoking a Taylor's frozen turbulence hypothesis, such time scales can often be expected to be proportional to model resolution. Based on this reasoning, the adjustment time–scale in the ECMWF convection parameterization closure is, indeed, set to proportional to the model resolution [71]. This argument can, in principle, be applied to any parameterization parameters: we can estimate the scale dependence of a given parameter based on a dimensional analysis. Importantly, we do not require any more sophisticated physical analysis here. The proportionality factor can be considered as a rather straightforward "tuning". A classical example is Smagorinksy's [139] eddy diffusion coefficient, which is designed to be proportional to the square of the model grid length. The pre-factor here is considered as "tuning" but the functional form is known beforehand. This is another example when we do not require extensive process studies (*cf.* Section 1.4): almost everything can be defined within a parameterization in a stand–alone manner under a good and careful theoretical construction.

A simple application of this idea for using dimensional analysis and scaling leads to a simple condition for turning off convection parameterization with increasing resolution. Convection could be characterized by a turn–over time scale and the criterion would be to turn off the convection parameterization when the turn–over time scale is longer than the minimum resolved time–scale. The latter would be estimated as a factor of few of the model time step with the exact factor depending on the model numerics. Since the former is proportional to the convection height, this condition would turn off parameterized convection first for the deepest clouds and gradually for shallower ones also as the model resolution increases. Here, again, we caution against a common custom of turning off a convection parameterization completely at a somewhat arbitrary model resolution, as already suggested in **T3.3**, **Q3.4.1**.

Clearly the best strategy in parameterization development would be to avoid an introduction of a scale–dependent parameter as much as possible, because it automatically eliminates a need for adjusting a scheme against the model resolution.

The second aspect to realize is that a scale gap is not a *prerequisite* for parameterizations. A separation between above–grid and subgrid scales is made rather in arbitrary manner (*cf.* **Q3.4.1**). In this respect, the best strategy for avoiding a double counting is to keep a consistency of a given parameterization with an original full system. For example, in order to avoid a double counting of energy–containing scales, a parameterization should contain a consistent energy cycle.

It is often anticipated that as a whole, resolved and parameterized convection are "communicating vessels" in a model. Thus, when parameterized convection is strong there is less intense and/or less likelihood of producing resolved convection and *vice versa*. Due to this tendency, the issue of double counting would not come out as a serious one most of time in operational experiences. However, this is true only if a model is well designed, and in fact, many models suffer from problems because they are not able to perform such a smooth transition between the "communicating vessels". This emphasizes the need for carefully constructing scale–independent physical schemes based on the principles outlined here. It, furthermore, reminds us the importance of constructing all physical schemes with due regard to generality (*cf.* Section 2.3.1) and in a self-consistent manner (*cf.* **T2.4**), as already emphasized, in order to avoid these operational difficulties.

Q3.4.3: What is the Degree of Complexity of Physics Required at a Given Horizontal Resolution?

Currently, various model sensitivities are discussed in a somewhat arbitrary manner: cloud physicists tend to focus on smaller scales in order to emphasize the sensitivities of model behavior to microphysical details, whereas dynamicists tend to focus on larger scales in order to emphasize the dynamical control of a given system. The ultimate question of sensitivities depends on the time and the space scales at which the model is intended to provide useful results. Such considerations of scale dependence must clearly be included in any sensitivity studies.

From a point of view of probability theory [140,141], this issue would be considered that of Occam's principle: if two physical schemes with different complexities provide us an equally good result (under a certain error measure), we should take a simpler one among the two (*cf.* link to **T4.2**).

Issues of physics complexity must also be considered in terms of the capacity of a given model for performing over a range of model resolutions. Some schemes may represent a well–behaved homogeneous behavior over a wide range of resolutions, avoiding brutal changes of forecast model structure and avoiding any parasitic manifestations, such as grid-point storms at the scales where convection must still be parameterized although it may partially be resolved. Such schemes would likely be able to be extended with additional physical complexity relatively easily in comparison with schemes that behave less well over the same variation of resolution.

Regardless of their complexity, any formulations for physical processes ultimately contain uncertainties (e.g., values of constant coefficients). Uncertainty of physical formulation leads to a growth of prediction uncertainty with time. Uncertainty growth can be well measured under a general tangent linear formulation which by constructing a linear perturbation equation along the original full model solution. It provides an exponential deviation rate from the original full solution with time as well as a preferred direction for the deviation (*i.e.*, a spatial pattern growing with time). This method also allows us to systematically examine feedback of a physical process to all the others. Note that the main question here is a sensitivity to the other physical processes (parameterized or not) by changing a one. In principle, the uncertainty growth estimated by a tangent linear method can be translated into a probability description by writing down the corresponding Liouville equation for a given tangent linear system. In these sensitivity–uncertainty analyses, uncertainties associated with physical parameters as well as those associated with an initial condition, observational uncertainties can be quantified.

2.4. Physics and Observations

This section examines various physical processes (notably cloud microphysics) important for convection as well as issues of observations.

T4.1: Review of Subgrid-Scale Microphysical Parameterizations

This assignment can be interpreted in two different ways: (i) review of microphysical parameterizations themselves (*i.e.*, phenomenological description of the microphysics); and, (ii) review of cloud microphysical treatments in convection parameterization. A review on the bin and bulk microphysical formulations has been developed [142] in response to the first issue. The second issue is dealt with by one of book chapters [143], and some relevant aspects are already discussed in **Q2.1.2**.

A special direction of investigation and interest is the effect of aerosols on the intensity of tropical cyclones. Khain *et al.* [144] and Lynn *et al.* [145] show that a model with bin-microphysics is able to predict the intensity of TC much better than current bulk-parameterization schemes. Furthermore, continental aerosols involved in the TC circulation during landfall decrease the intensity of TC to the same extent as the sea surface temperature cooling caused by the TC–ocean interaction. These studies indicate the existence of an important hail–related mechanism that affects TC intensity that should be taken into account to improve the skill of the TC forecasts.

The present Action has also developed an innovative new theory for time-dependent freezing [146,147]. This work highlights another key advantage of bin microphysics schemes: representing particle properties that have strong size-dependence. Wet growth of hail happens only when a critical size is exceeded, and particles that become wet carry their liquid during size-dependent sedimentation. A new theory for such freezing is developed for bin-microphysical schemes. The theory encompasses wet growth of hail, graupel and freezing drops.

The new algorithm has been implemented into the Hebrew University Cloud Model (HUCM) and mid-latitudinal hail storms have been simulated under different aerosol conditions. It is shown that a large hail width diameter of several centimeters forms only in the case of high (continental) aerosol loading [148]. It is also shown that hail increases precipitation efficiency leading to an increase in surface precipitation with an increase in the aerosol concentration.

For the first time all the parameters measured by Doppler polarimetric radar have been evaluated. These parameters have been calculated according to their definitions using size distribution functions of different hydrometeors in HUCM. A long-standing problem of the formation of so–called columns of differential reflectivity Zdr is solved by Kumjian [149] as a result. High correlations are found between Zdr on the one hand, and hail mass and size on the other [150]. This finding opens a way to improve the short–range forecast of hail, its size, and the intensity of hail shafts.

2.4.1. Further Processes to be Incorporated into Convection Parameterizations

In addition to the cloud microphysics, the following processes may be considered to be important for convective dynamics as well as convection parameterization. However, again, we emphasize here the difference between the two issues (*cf.* Sections 1.3 and 1.4): importance in convective dynamics and that in convection parameterization. The following discussions are also developed under an emphasis of this distinction.

2.4.1.1. Downdrafts

Downdrafts have long been identified as a key process in convective dynamics [151–153]. From a theoretical point of view, the importance of downdrafts has been addressed in the context of tropical–cyclone formation [154] as well as that of Madden–Julian oscillations [155]. Although the majority of current operational mass–flux convection parameterizations do include convective downdrafts in one way or another (e.g., [156–159]), they are implemented rather in an *ad hoc* manner. The downdraft formulation must be more carefully constructed from a more general principle , e.g., SCA (*cf.* **Q1.8**).

However, the real importance of downdrafts in convection parameterization must also be carefully re-assessed. The thermodynamic role of convection is to dry and cool the boundary layer by a vertical transport process. The updraft and downdraft essentially perform the same function by transporting thermodynamic anomalies with opposite signs in opposite directions. In the above listed theoretical investigations, it is interesting to note that the downdraft strength is measured in terms of precipitation efficiency. This further suggests that the same effect may be achieved by simply re–distributing the downdraft effect into deep updraft and shallow–convective mixing. Thus, sensitivities of downdrafts on model behavior demonstrated by these studies do not necessarily demonstrate the true importance the downdraft representation in convection parameterization.

Refer to a book chapter [160] on the state-of-the-art of the downdraft parameterization, including the issues of saturated and unsaturated downdrafts. Also note a link of the issues to the next section.

2.4.1.2. Cold Pools

The cold pool in the boundary layer is often considered a major triggering mechanism for convection. Observations suggest that cold pool-generated convective cells occur for shallow maritime convection [161,162], maritime deep convection [163–165] and continental deep convection (e.g., [166–169]). Moreover, numerical studies appear to suggest that cold pools promote the organization of clouds into larger structures and thereby aid the transition from shallow to deep convection [170–172]: but see [9,42]). A cold–pool parameterization coupled with convection is already proposed by Grandpeix and Lafore [173], and Rio *et al.* [30], although we should view it with some caution [174].

However, the evidence for cool–pool triggering of convection remains somewhat circumstantial, and a clear chain of cause and effect has never been identified. For example, Boing *et al.* [171] attempt sensitivity study by performing LES experiments with the four different set–ups with the aim of elucidating a feedback loop. However, this article fails to list the processes actually involved in a feedback loop. Though a careful Lagrangian trajectory analysis is performed, they fail to identify whether the Lagrangian particle has originated from a cold pool or not.

From a point of view of the convective energy cycle already discussed in **Q1.2.1** (see also [9,42]), deep convection is induced from shallow convection by the tendency of the latter continuously destabilizing its environment by evaporative cooling of non–precipitating water as it detrains. More precisely, this process increases both the available potential energy and the cloud work function for deep convection, and ultimately leads to an induction of deep convection, as manifested by a sudden increase of its kinetic energy. In a more realistic situation, the evaporative cooling may more directly induce convective downdrafts, which may immediately generate cold pools underneath. These transformations are clearly associated with an induction of kinetic energy from available potential energy. However, the existing literature does not tell us whether those pre-existing kinetic energy associated either with downdraft or cold pool is transformed into deep–convective kinetic energy, for example, by pressure force, or alternatively, an extra process is involved in order for a cold pool to trigger convection. The CRM can be used for diagnosing these energy–cycle processes more precisely. An outline for such a method is described by Figure 5 and associated discussions in Yano *et al.* [101]. However, unfortunately, this methodology has never been applied in detail even by the original authors.

We may even point out that the concept of a "trigger" is never clearly defined in literature, but always referred to in a phenomenological manner, even in an allegorical sense. The same notion is never found either in fluid mechanics or turbulence studies. Recall that we have already emphasized that the notion of a "trigger" is fundamentally at odds with the basic formulation of the mass–flux parameterization (*cf.* Section 2.1.1, **T1.1**, **Q1.2.1**). As also emphasized in Section 2.1.1, in order to introduce a trigger process into a parameterization, a radical modification of its formulation is required.

2.4.1.3. Topography

Topography can often help to induce convection by a forced lifting of horizontal winds (*cf.* [175,176]). Thus, subgrid–scale topography is likely to play an important role in triggering convection. Studies on subgrid–scale topographic trigger of convection as well as assessment of the possible need for incorporating this into a parameterization are still much missing.

2.4.2. Link to the Downscaling Problem

As already emphasized in several places, the goal of parameterization is to provide a grid–box averaged feedback of a subgrid–scale process to a large–scale model. Thus, any subgrid–scale details themselves are beyond a scope of a parameterization problem. However, in some applications, these subgrid–scale details often become their own particular interests. A particularly important example is a prediction of local extreme rainfall, that typically happens at a scale much smaller than a model grid size. A procedure for obtaining such subgrid–scale details is called downscaling. A link between parameterization and downscaling is emphasized by Yano [177], and much coordinated efforts on these two problems are awaited.

T4.2: Proposal and Recommendation on Observational Validations

A review on observational validation of precipitation is found as one of the book chapters [178]. However, unfortunately, current validation efforts are strongly application oriented and weak in theoretical, mathematical basis. Especially, the current methods are not able to identify a missing physical process that has led to a failed forecast. The use of wavelet analyses, for example, could help to overcome this by establishing a link between forecast validations and model physical processes. The importance of the precipitation–forecast validation is also strongly linked to issues associated with the singular nature of precipitation statistics (strongly departing from Gaussian, and even from log-Gaussian against a common belief), leading to particular importance of investigating extreme statistics (*cf.* Section 2.4.2).

In the longer term, the need for probabilistic quantifications of the forecast should be emphasized, as already suggested at several places (*cf.* **Q2.1.2**, **Q3.4.3**). We especially refer to Jaynes [140] and Gregory [141] for the basics of the probability as an objective measure of uncertainties. From the point of view of probability theory, the goal of the model verification would be to reduce the model uncertainties by objectively examining the model errors. In order to make such a procedure useful and effective, forecast errors and model uncertainties must be linked together in a direct and quantitative manner. Unfortunately, many of the statistical methods found in general literature are not satisfactory for this purpose. The Bayesian principle (*op. cit.*) is rather an exception that can provide such a direct link so that from a given forecast error, an uncertainty associated with a particular parameter in parameterization, for example, can objectively and quantitatively estimated. The principle also tells us that ensemble, sample space, randomization, *etc.* as typically employed in statistical methods are not indispensable ingredients for uncertainty estimates, although they may be useful. Though there are already many applications of Bayesian principle to atmospheric science, the capacity of Bayesian for linking between the statistical errors and physical processes has not much been explored (*cf.* [179,180]). See also **Q3.4.3** for a link to issues of required model complexity and uncertainties.

T4.3: Proposal and Recommendation for a Parameterization with Unified Physics

The issue of "unified physics" is often raised in existing reviews on the subgrid–scale parameterization problem. The best example would be Arakawa [3]. To directly quote from his abstract: "for future climate models the scope of the problem must be drastically expanded from 'cumulus parameterization' to 'unified cloud parameterization,' or even to 'unified model physics.' This is an extremely challenging task both intellectually and computationally, ..." However, we have to immediately realize that, at the most fundamental level, there is no need for unifying any physics in the very context of atmospheric science.

The best historical example for an issue of unification of physics is the one between mechanics and electromagnetism encountered towards the end of the 19th century. The system of mechanics is invariant under the Galilean transform, whereas the system of electromagnetism is invariant under the Lorentz transform. Thus these two systems were not compatible each other. This led to a discovery (or more precisely a proposal) of relativity by Einstein, that unified the physics.

However, there is no analogous issue of unification in atmospheric science. Our model construction starts from a single physics. Of course, this is not to say that all the physics are already known. That is clearly not the case, particularly for cloud microphysics. However, the issue of "unified physics" in atmospheric science arises not through any apparent contradiction in the basic physics but only after a model construction begins.

To develop a numerical model of the atmosphere, often, a set of people are assigned separately for the development of different physical schemes: one for clouds, another for convection, a third for boundary layer processes, *etc.* Often this is required because the development of each aspect needs intensive concentration of work. This also leads to separate development of code for different "physics". However, a complete separation of efforts does not work ultimately, because, for example, clouds are often associated with convection, and convection with clouds. The treatment of clouds and convection within the boundary layer faces a similar issues: should they be treated as a part of a boundary-layer scheme simply because they reside in the boundary layer?

It transpires that the issue of unification of the physics only happens in retrospect, and only as a result of uncoordinated efforts of physical parameterization development. If everything were developed under a single formulation, such a need should never arise afterwards. In this very respect, the main issue is more of a matter of the organization of model development rather than a real scientific issue (*cf.* Section 2.3.1). In order to follow such a method of course a certain general methodology is required. That has been the main purpose of the present Action. See **T2.4**, Section 2.3.1, and [101] for further discussions.

It could be tempting to add the existing parameterization codes together in consistent manner from a practical point of view. However, such an approach could easily turn out be to less practical in the long term. One may add one more dictum to a list already given in Section 2.3.1: never go backwards. As already suggested in Section 2.4.1, for example, it is something of a historical mistake that downdrafts are introduced into the mass–flux formulation in an ad hoc manner.

As already emphasized in Section 2.3.1, it is imperative to re–derive all the schemes from basic principles in order to establish unified physics, checking consistency of each hypothesis and approximation. This may sound a painful process. However, this is what every researcher is expected to do whenever he or she tries to use a certain physical scheme. In the end, if the original development has been done relatively well, the resulting modifications to a code could also be relatively modest. In order to establish such consistency in combination with the use of a mass–flux parameterization, SCA and its further relaxations would become an important guiding principle, as already suggested in **T2.4**. By relaxing SCA, it is straightforward to take into account of a certain distribution over a particular subgrid–scale component (segment); especially, over the environment. Such an idea (EDMF) is first introduced by Soares *et al.* [54], and an SCA procedure can derive such a formulation in a more self–consistent manner (*cf.* **T1.5**).

All of the non–mass flux based parameterizations, such as eddy transport, must be handled in this manner for consistency. Note that these non–mass–flux–based parameterizations may be introduced into different subgrid–scale components, for example, into the environment and convection separately (*cf.* **T1.5**). In order to maintain overall consistency, a fractional area occupied by a given subgrid–scale component must explicitly be added to the formulation. Note that this pre–factor is usually neglected in standard non–mass–flux parameterizations assuming that convection occupies only an asymptotically vanishing fraction. In the high resolution limit, this assumption becomes no longer true, as already discussed in **T3.2** and **T3.3**.

Importantly, such a pre–factor can easily be added to an existing code without changing its whole structure. This is just an example to demonstrate how the consistency of physics can be re–established by starting from the first principle of derivations, but without changing the whole code structure: use your pencil and paper carefully all the way, which is a totally different task than coding.

Q4.3.1: How can a Microphysical Formulation (Which is by Itself a Parameterization) be Made Resolution Dependent?

First of all, as a minor correction, according to the terminology already introduced in Section 1.1, the microphysical formulation is *not* by itself a parameterization, although it is most of time *phenomenologically* formulated. These phenomenological microphysical descriptions are defined at each "macroscopic" point. In this context a macroscopic point is defined as having a spatial extent large enough so that enough molecules are contained therein but also small enough so that spatial inhomogeneities generated by turbulent atmospheric flows are not perceptible over this scale. Such a scale is roughly estimated as a micrometer scale.

A typical model resolution is clearly much larger that this scale, and thus a spatial average must be applied to the phenomenological microphysical description developed for a single macroscopic point. Averaging leads to various Reynolds' stress–like and nonlinear cross terms, which cannot be described in terms of resolved–scale variables in any obvious manner.

Little explicit investigation of this issue has been performed so far. Especially, a mathematical theory is required in order to estimate cross correlation terms under an expected distribution of the variables in concern.

Q4.3.2: Can Detailed Microphysics with Its Sensitivity to Environmental Aerosols be Incorporated into a Mass-Flux Convection Parameterization? Are the Current Approaches Self-Consistent of Not? If Not, How can It be Achieved?

The response of convective clouds and precipitation to aerosol perturbations is intricate and depends strongly on the thermodynamic environment [181], but the response is potentially very large [182], thus an adequate representation of the effects in numerical weather prediction would be desirable, and probably more so for climate projections. The microphysics currently applied within convective parameterizations has not much improved since e.g., [157].

The primary interactions between aerosols and convection are the wet scavenging of aerosols by precipitation, and the co-variation of wet aerosol mass and aerosol optical depth with cloud properties in the humid environment of clouds [183]. A very simplified microphysical representation of the effect of aerosols on convective precipitation would be to use a "critical effective radius", or a threshold in effective droplet size before the onset of rain [184]. For this reason, the size of droplets that grow in the updraft is taken instead of the height above cloud base as in [157] in order to determine the convective precipitation rate. This implementation has been tested in the ECHAM general circulation model by Mewes [185]. A large effect is found probably because of a missing wet scavenging in this model version. The actual convective invigoration hypothesis [182] cannot be tested in a convection parameterization, though, until freezing and ice microphysics are implemented.

2.4.3. How Can Observations Be Used for Convection Parameterization Studies?

Observations (and measurements) are fundamental to science. However, further observations are not necessarily useful per se, and they require a context in which it has been established what we have to observe (or measure) for a given purpose, including specifications of the necessary temporal and spatial resolution as well as the accuracy of the measurements.

Parameterization development and evaluation may ideally take a two-step approach [186]. Insights into new processes and initial parameterization formulation should be guided by theory and process-level observations (laboratory experiments and field studies). The latter may be substituted by LES/CRM–based modelling, if suitable observations are unavailable. However, once implemented, further testing and evaluation are required in order to ensure that the parameterization works satisfactorily for all weather situations and at the scales used in a given model. Satellite observations are probably the most valuable source of information for the latter purpose, since they offer a large range of parameters over comparatively long time series and at a very large to global coverage. The A-train satellites may be noted as a particular example. In order to facilitate such comparisons, "satellite simulators" have been developed, which emulate satellite retrievals by making use of model information for the subgrid-scale variabilities of, for example, clouds leading to statistics summarizing the model performances [187–189]. A large range of methodologies has also been developed in terms of process-oriented metrics, e.g., for investigating the life cycle of cirrus from convective detrainment [190], and for elucidating the details of microphysical processes [191]. In addition to those techniques focusing on individual parameterizations, data assimilation techniques can also be exploited as a means of objectively adjusting convection parameters and learning about parameter choices and parameterizations [192].

However, fundamental limits of current satellite measurements must be recognized. Most satellites measure only cloud optical properties (either in visible or invisible range of light). These quantities, unfortunately, do not provide much useful direct information about the dynamical convective processes (e.g., vertical velocity).

In reviewing the closure problem [20], the difficulties were striking for the apparently simple task of identifying convection objectively from observations. Most of the observational analyses that we have reviewed use the precipitation rate as a measure of convection. Although this could be, partially, acceptable over the tropics, such a measure is no longer useful enough over the midlatitudes, where much of the precipitation originates from synoptic scale processes. It may be needless to emphasize that satellite images (such as outgoing wave radiation) are even less reliable as a direct measure of convection.

After a long process of discussions, we finally identify lightning data as, probably, the best measure of convection currently available at a routine level. In our best knowledge, lightning happens only in association with strong vertical motions and extensive ice, so that the cloud must be high enough and dynamically very active. A fair objection to this methodology would be the fact that we would still miss some convection under this strategy, and that other factors (e.g., aerosol conditions, degree of glaciation of convection) may also influence the lightning even if the vertical motions are unchanged. We also have to keep in mind that this is only a qualitative measure without giving any specific quantification such as convective vertical velocity. However, importantly, when there is lightning, this is a sure sign that there is convection.

Satellite lightening data from NASA's OTD (Optical Transient Detector) mission exists for the period of April 1995–March 2000 (thunder.msfc.nasa.gov). Lightning data from a ground-based lightning–location system is also available over Europe. Such a network has a high detection efficiency (70%–90%) and location accuracy (<1 km). EUropean Cooperation for Lightning Detection (EUCLID: euclid.org), consisting of 140 sensors in 19 countries, is currently the most comprehensive network over Europe organized under a collaboration among national lightning detecting networks. Based on this measurement network, we are planning to perform systematic correlation analysis between lightning frequencies and other physical variables (column-integrated water, CAPE, etc.). This project would be considered an important outcome from the present COST Action.

As a whole, we emphasize the importance of identifying the key variables as well as processes in order to analyze convection for the development, verification, and validation of parameterizations. Very ironically, in this very respect, the traditional Q1 and Q2 analysis based on a conventional sounding network can be considered as the most powerful observational tool for convection parameterization studies. The reason is very simple: that these are the outputs we need from a parameterization, and thus must be verified observationally.

3. Conclusions: Retrospective and Perspective

The Action is to identify the closure and the entrainment–detrainment problems as the two highest priorities for convection parameterization studies under the mass–flux formulation. The conclusion is rather obvious in retrospect: closure and entrainment–detrainment are the two major cornerstones in mass–flux convection parameterization formulation. Unless these two problems are solved satisfactory, the operational convection parameterization under mass–flux formulation would never work satisfactory.

It is rather surprising to realize that it took us four years of reflections in order to reach this very simple basic conclusion collectively. The original MoU, prepared by the members by extending editing, has even failed to single out these two basic issues. MoU has wrongly identified the "convective triggering conditions" as one of key elements of mass–flux formulation along with closure and entrainment–detrainment (Section B.2, *cf.* Section 2.1.1, **T1.1**). MoU also provides a false anticipation (the 4th secondary objective: *cf.* Section 1.5) that extensive process studies by CRM and LES would *by themselves* automatically lead to improvements of parameterization. In the end of the present Action, we openly admit our misjudgment. Though the value of the process studies should hardly diminish in its own right, they do not serve for a purpose of parameterization improvements *by themselves* automatically unless we approach to the problem from a good understanding of the latter. This point is already extensively discussed in Sections 1.3, 1.4, 2.4.1, **Q3.4.3** and re–iterated again below.

The intensive theoretical reflections on the subgrid–scale parameterization problem over these four years have been a very unique exercise. We can safely claim that all our meetings have been great success with great satisfactions of the participants. However, this rather reflects a sad fact that such in–depth discussions on the issues from scientific theoretical perspectives (and just simply making any logic of an argument straight) are rarely organized nowadays with dominance of approaches seeking technological solutions with massive modeling and remote sensing data. At the more basic level, the present Action gives a lesson on importance to just sit and reflect: the current scientific culture needs to be much changed towards this direction.

The present report summarizes the achievements which have been made during the Action ES0905. Those results would have been possibly only under a support of COST, as suggested in the last paragraph. A healthy scientific environment in our discipline (as in many others) requires healthy respect and balance between observational, modeling and theoretical investigations. An imbalanced situation ultimately hinders progress. The present Action has been motivated from a sense that theoretical investigations are currently under–weighted. This sense has been reinforced during the Action itself, which has shown promise that even relatively modest efforts towards redressing the balance could prove extremely fruitful.

However, setting the two identified priorities, closure and entrainment–detrainment, actually into forefront of research for coming years is already challenging. We first of all need to overcome the basic strategies of participating agencies already so hard–wired differently before we can launch such an ambitious project. The current economic crisis rectifies this general tendency further with even more emphasis on technological renovations rather than fundamental research at the EU level. Budget cuts at a level of individual institution is more than often associated with a more focus on short term deliverables and products, rather than more fundamental research. Such short–term focused (and short–sighted) strategies are likely to lead to depletion of creative real innovative ideas in longer terms.

Improving parameterization is a long and difficult path, due to the many feedbacks within numerical models for weather and climate. True improvements are possible only in association with fundamental research. Though in short terms quick dirty fixes are possible, they would be likely to lead to long–term deterioration of true quality of models. This whole situation has already begun to deplete future visions for research: in spite of the very logical basic conclusion, majority of participating researchers do not feel ready to focus themselves on these basic issues. Clearly we are short of specific plans for tackling them. On the closure problem, a few possibilities are listed, but we are far from reaching any consensus. The situation with entrainment and detrainment is even worse. Though we identify couple of elements to consider, there is no identified line for further theoretical investigations.

Less is even said about perspectives for developing a more fundamental research from a turbulence perspective. It is long stated that convection parameterization is a statistical description (*cf.* [49]). However, a statistical mechanics for describing ensemble convection system is still to be emerged even after 40 years. Here, we even face a difficulty for obtaining a funding towards this goal at an European level. Under this perspective, the phenomenologically–oriented process study would become the highest priority for years to come, though fundamental limitations of this approach must well be kept in mind. The best lesson to learn is an improvement of QN into EDQN as discussed in Sections 1.3 and 1.4. It is easy to criticize the QN model from a phenomenological perspective. Indeed, it neglects various phenomenologically well–known aspects of turbulence: intermittency, inhomogeneity, *etc*. In the same token, it is easy to criticize the current convection parameterization based on lack of various phenomenological elements: lack of life cycle, coupling with boundary layer, mesoscale organization, *etc*.

However, making these statements by itself does not improve a parameterization in any manner. Further investigations on these processes themselves do not contribute to an improvement of a parameterization in any direct manner, either. In order to move to this next step, a strategy (or more precisely a formulation) for implementing those processes must exist. Even first required is a clear demonstration that lack of these processes is actually causing a problem within a given parameterization. Recall that a parameterization may run satisfactory by still missing various processes what we may consider to be crucial. For example, Yano and Lane [193] suggest that the wind shear may not be crucial for thermodynamic parameterization of convection, although it plays a critical role in organizing convection in mesoscale. The goal of the parameterization is just to get a grid–box averaged feedback of a given whole process (not each element) correct. No more detail counts by itself.

In this respect, the evolution of QN into EDQN is very instructive. The study of Orszag in 1970 [6] on the QN system identified that the real problem of QN is not lack of intermittency or inhomogeneity, but simply due to an explosive tendency in growth of skewness in this system. This lesson tells us that it is far more important to examine the behavior of the actual parameterization concerned, rather than performing extensive phenomenological process studies. Turbulence studies also suggest that what appears to be phenomenologically important may not be at all crucial for purely describing large–scale feedbacks. For example, Tobias and Marston [194] show that even a simple statistical model truncated at the second order of cumulant (which is a variant of moment adopted in the QN model) can reproduce realistic multiple jets for planetary atmospheres for a realistic parameter range, in spite of the fact that this model neglects many intrinsic characteristics of the turbulence, as the case for QN. Here, we see a strong need for a shift from the process studies to the formulation studies in order to tackle with the convection parameterization problem in a more direct manner. The present report has suggested in various places how such formulation studies are possible more specifically.

As the present Action has identified, the parameterization problem (as for any other scientific problems) is fundamentally even *ontological* [4]. The present situation is like blind people touching different parts of an elephant (the trunk, a leg, the nose, ...), and arguing harshly over the true nature of the elephant without realizing that they are only touching a part of it. We scientists are, unfortunately, not far from those blind people so long as the parameterization problem is concerned. Thus, the last recommendation in concluding the present Action is to create a permanent organization for playing such an ontological role in the parameterization problem, possibly along with the other fundamental problems in atmospheric modeling. Here, the ontology should not mean pure philosophical studies. Rather, it should be a way for examining the whole structure of a given problem from a theoretical perspective in order to avoid myopic tendencies of research. The present Action has played this role for last four years, and it is time to pass this responsibility to a permanent entity.

Acknowledgments

The present report has emerged through the meetings as well as many informal discussions face to face and over e-mails for the whole duration of the COST Action ES0905. The lead author sincerely thanks to all the COST participants for this reason. Especially, in preparing the manuscript, Florin Spineanu, Linda Schlemmer, Alexander Khain have provided text segments. Parts of the manuscript are carefully proof read by Peter Bechtold, Alexander Bihlo, Elsa Cardoso, Alan L. M. Grant. Marja Bister and Sandra Turner have provided specific comments. Wim de Rooy has gone through the whole manuscript several times for critical reading and with suggestions for elaborations. Discussions on turbulence modelling with Steve M. Tobias and a reference on TRMM data from Steve Krueger are also acknowledged. The image for Figure 1 is provided by Szymon Malinowski. DM's contribution to the present paper is limited to the WG3-related issues.

Author Contributions

The lead author took a responsibility of putting all the contributions together into a single coherent text. He also took a final responsibility of edit. The other authors contributed text segments of varying lengths, read the texts at various drafting stages, and contributed comments, critics, modifications of the text, as well for editing. DM's contribution is limited to Section 2.3.

Conflicts of Interest

The authors declare no conflicts of interest.

Appendix: Derivation of the Result for T3.2

In this Appendix, we outline the results verbally stated in **T3.2**. For the basics of the mass–flux formulation, readers are advised to refer to, for example, [12]. For simplicity, we take a bulk case that the subgrid processes only consist of environment and convection with the subscripts e and c, respectively. Thus, any physical variable, φ, at at any given gird point is decomposed as

$$\varphi = \sigma_e \varphi_e + \sigma_c \varphi_c \qquad (A1)$$

where σ_e and σ_c are the fractional areas occupied by environment and convection, respectively. Clearly,

$$\sigma_e + \sigma_c = 1 \qquad (A2)$$

With the help of Equation (A2), Equation (A1) may be re-written as

$$\varphi = \varphi_e + \sigma_c(\varphi_c - \varphi_e) \qquad (A3)$$

Under the standard mass–flux formulation (*cf.* Section 7 in [12], the asymptotic limit $\sigma_c \to 0$ is taken. As a result, the grid–point value may be approximated by the environmental value in this limit

$$\varphi \to \varphi_e \qquad (A4)$$

as seen by referring to Equation (A3).

The goal of this Appendix is to infer the corrections due to finiteness of σ_c. For this purpose, we expand physical variables by σ_c. For example, the grid–point value is expanded as

$$\varphi = \varphi^{(0)} + \sigma_c \varphi^{(1)} + \cdots \qquad (A5)$$

where we find

$$\varphi^{(0)} = \varphi_e \qquad (A6)$$

$$\varphi^{(1)} = \varphi_e^{(0)} - \varphi_e \qquad (A7)$$

On the other hand, we find that the environmental value, φ_e, can be most conveniently integrated directly in time without expanding in σ_c by the equation:

$$\sigma_e \frac{\partial}{\partial t} \varphi_e + \frac{D}{\rho}(\varphi_e - \varphi_c) + \sigma_e w_e \frac{\partial}{\partial z} \varphi_e + \sigma_e \mathbf{u}_e \cdot \bar{\nabla}\varphi_e = \sigma_e F_e \qquad (A8)$$

which is given as Equation (6.7) in [12]. Here, w_e and $\mathbf{u}_e$ are the environmental vertical and horizontal velocities, and F is the forcing term for a given physical variable. The time integration of Equation (A8) is essentially equivalent to that for the grid–point equation under the standard approximation (*cf.* Equation (7.8) in [12]), but with the grid–point value replaced by the environmental value.

The equivalent equation for the convective component is given by Equation (6.4) of Yano [12]:

$$\sigma_c \frac{\partial}{\partial t} \varphi_c + \frac{E}{\rho}(\varphi_c - \varphi_e) + \sigma_c w_c \frac{\partial}{\partial z} \varphi_c + \sigma_c \mathbf{u}_c \cdot \bar{\nabla}\varphi_c = \sigma_c F_c \qquad (A9)$$

The basic idea here is, instead of solving the above prognostic equation directly, to solve the problem diagnostically by performing an expansion in σ_c. Here, we expect that the forcing, $\sigma_c F_c$, in convective scale is a leading–order quantity, thus we expand this term as

$$\sigma_c F_c = F_c^{(0)} + \sigma_c F_c^{(1)} + \cdots \tag{A10}$$

To leading order of expansion, we essentially recover an expression for the standard mass–flux formulation already given by Equation (8) but with forcing:

$$M \frac{\partial \varphi_c^{(0)}}{\partial z} = -E(\varphi_c^{(0)} - \varphi_e) + \rho F_c^{(0)} \tag{A11}$$

This equation can be solved by vertically integrating the right hand side.

To the order σ_c, we obtain a correction equation for the above as

$$M \frac{\partial \varphi_c^{(1)}}{\partial z} + E \varphi_c^{(1)} - \rho F_c^{(1)} = -\frac{\partial}{\partial t} \varphi_c^{(0)} \tag{A12}$$

All the terms gathered in the left hand side is equivalent to the leading order problem (A11) except for that is for $\varphi_c^{(1)}$. Furthermore, they are modified by a temporal tendency, $\partial \varphi_c^{(0)}/\partial t$, already known from the leading order, given in the right hand side. It immediately transpires that this diagnostic procedure is nothing other than performing a time integral of the whole solution under a special "implicit" formula, thus the statement in **T3.2** follows.

References

1. Plant, R.S.; Yano, J.-I. *Parameterization of Atmospheric Convection*; Imperial College Press: London, UK, 2014; Volumes I and II, in press.
2. McFarlane, N. Parameterizations: Representing key processes in climate models without resolving them. *WIREs Clim. Chang.* **2011**, *2*, 482–497.
3. Arakawa, A. The cumulus parameterization problem: Past, present, and future. *J. Clim.* **2004**, *17*, 2493–2525.
4. Yano, J.-I.; Vlad, M.; Derbyshire, S.H.; Geleyn, J.-F.; Kober, K. Generalization, consistency, and unification in the parameterization problem. *Bull. Amer. Meteor. Soc.* **2014**, *95*, 619–622.
5. Lesieur, M. *Turbulence in Fluids*; Martinus Nijhoff Publisher: Dordrecht, Netherlands, 1987.
6. Orszag, S.A. Analytical theories of turbulence. *J. Fluid Mech.* **1970**, *41*, 363–386.
7. Guichard, F.; Petch, J.C.; Redelsperger, J.L.; Bechtold, P.; Chaboureau, J.-P.; Cheinet, S.; Grabowski, W.; Grenier, H.; Jones, C.G.; Köhler, M.; *et al.* Modelling the diurnal cycle of deep precipitating convection over land with cloud-resolving models and single-column models. *Quart. J. R. Meteor. Soc.* **2004**, *604*, 3139–3172.
8. Lenderink, G.; Siebesma, A.P.; Cheinet, S.; Irons, S.; Jones, C.G.; Marquet, P.; Müller, F.; Olmeda, D.; Calvo, J.; Sánchez, E.; *et al.* The diurnal cycle of shallow cumulus clouds over land: A single-column model intercomparison study. *Quart. J. R. Meteor. Soc.* **2004**, *130*, 3339–3364.
9. Yano, J.-I.; Plant, R.S. Coupling of shallow and deep convection: A key missing element in atmospheric modelling. *J. Atmos. Sci.* **2012**, *69*, 3463–3470.

10. Bechtold, P.; Semane, N.; Lopez, P.; Chaboureau, J.-P.; Beljaars, A.; Bormann, N. Representing equilibrium and non-equilibrium convection in large-scale models. *J. Atmos. Sci.* **2014**, *71*, 734–753.

11. Yano, J.-I.; Graf, H.-F.; Spineanu, F. Theoretical and operational implications of atmospheric convective organization. *Bull. Am. Meteor. Soc.* **2012**, *93*, ES39–ES41

12. Yano, J.-I. Formulation structure of the mass–flux convection parameterization. *Dyn. Atmos. Ocean* **2014**, *67*, 1–28.

13. Yano, J.-I. Formulation of the mass–flux convection parameterization. In *Parameterization of Atmospheric Convection*; Plant, R.S., Yano, J.I., Eds.; Imperial College Press: London, UK, 2014; Volume I, in press.

14. Marquet, P. Definition of a moist entropy potential temperature: Application to FIRE-I data flights. *Quart. J. R. Meteor. Soc.* **2011**, *137*, 768–791.

15. Marquet, P. On the computation of moist-air specific thermal enthalpy. *Quart. J. R. Meteor. Soc.* **2014**, doi:10.1002/qj.2335.

16. Marquet, P.; Geleyn, J.-F. Formulations of moist thermodynamics for atmospheric modelling. In *Parameterization of Atmospheric Convection*; Plant, R.S., Yano, J.I., Eds.; Imperial College Press: London, UK, 2014; Volume II, in press.

17. Donner, L.J. A cumulus parameterization including mass fluxes, vertical momentum dynamics, and mesoscale effects. *J. Atmos. Sci.* **1993**, *50*, 889–906.

18. Yano, J.-I. Convective vertical velocity. In *Parameterization of Atmospheric Convection*; Plant, R.S., Yano, J.I., Eds.; Imperial College Press: London, UK, 2014; Volume I, in press.

19. Yano, J.-I. Interactive comment on "Simulating deep convection with a shallow convection scheme" by Hohenegger, C., and Bretherton, C. S., On PBL-based closure. *Atmos. Chem. Phys. Discuss.* **2011**, *11*, C2411–C2425.

20. Yano, J.I.; Bister, M.; Fuchs, Z.; Gerard, L.; Phillips, V.; Barkidija, S.; Piriou, J.M. Phenomenology of convection-parameterization closure. *Atmos. Phys. Chem.* **2013**, *13*, 4111–4131.

21. Del Genio, A.D.; Wu, J. The role of entrainment in the diurnal cycle of continental convection. *J. Clim.* **2010**, *23*, 2722–2738.

22. Stratton, R.A.; Stirling, A. Improving the diurnal cycle of convection in GCMs. *Quart. J. R. Meteor. Soc.* **2012**, *138*, 1121–1134.

23. Raymond, D.J. Regulation of moist convection over the warm tropical oceans. *J. Atmos. Sci.* **1995**, *52*, 3945–3959.

24. Zhang, G.J. Convective quasi-equilibrium in midlatitude continental environment and its effect on convective parameterization. *J. Geophys. Res.* **2002**, doi:10.1029/2001JD001005.

25. Zhang, G.J. The concept of convective quasi–equilibrium in the tropical western Pacific: Comparison with midlatitude continental environment. *J. Geophys. Res.* **2003**, doi:10.1029/2003JD003520.

26. Mapes, B.E. Convective inhibition, subgrid-scale triggering energy, and stratiform instability in a toy tropical wave model. *J. Atmos. Sci.* **2000**, *57*, 1515–1535.

27. Bretherton, C.S.; McCaa, J.R.; Grenier, H. A new parameterization for shallow cumulus convection and its application to marine subtropical cloud-topped boundary layers. Part I: Description and 1D results. *Mon. Weather Rev.* **2004**, *132*, 864–882.

28. Hohenegger, C.; Bretherton, C.S. Simulating deep convection with a shallow convection scheme. *Atmos. Chem. Phys.* **2011**, *11*, 10389–10406

29. Donner, L.J.; Phillips, V.T. Boundary layer control on convective available potential energy: Implications for cumulus parameterization. *J. Geophys. Res.* **2003**, doi:10.1029/2003JD003773.

30. Rio, C.; Hourdin, F.; Grandpeix, J.-Y.; Lafore, J.-P. Shifting the diurnal cycle of parameterized deep convection over land. *Geophys. Res. Lett.* **2009**, doi:10.1029/2008GL036779.

31. Rio, C.; Hourdin, F.; Grandpeix, J.-Y.; Hourdin, H.; Guichard, F.; Couvreux, F.; Lafore, J.-P.; Fridlind, A.; Mrowiec, A.; Roehrig, R.; *et al.* Control of deep convection by sub-cloud lifting processes: The ALP closure in the LMDD5B general circulation model. *Clim. Dyn.* **2012**, *40*, 2271–2292.

32. Mapes, B.; Neale, R. Parameterizing convective organization to escape the entrainment dilemma. *J. Adv. Model. Earth Syst.* **2011**, doi:10/1029/211MS00042.

33. Birch, C.E.; Parker, D.J.; O'Leary, A.; Marsham, J.H.; Taylor, C.M.; Harris, P.P.; Lister, G.M.S. Impact of soil moisture and convectively generated waves on the initiation of a West African mesoscale convective system. *Quart. J. R. Meteor. Soc.* **2013**, *139*, 1712–1730.

34. Bechtold, P. ECMWF, Reading, UK. Personal communication, 2014.

35. Semane, N. ECMWF, Reading, UK. Personal communication, 2013.

36. De Rooy, W. KNMI, De Bilt, the Netherlands. Personal communication, 2014.

37. Arakawa, A.; Schubert, W.H. Interaction of a cumulus cloud ensemble with the large-scale environment, pt. I. *J. Atmos. Sci.* **1974**, *31*, 674–701.

38. Yano, J.-I.; Plant, R.S. Convective quasi-equilibrium. *Rev. Geophys.* **2012**, *50*, RG4004.

39. Leith, C.E. Nonlinear normal mode initialization and quasi–geostrophic theory. *J. Atmos. Sci.* **1980**, *37*, 958–968.

40. Davies, L.; Plant, R.S.; Derbyshire, S.H. Departures from convective equilibrium with a rapidly-varying forcing. *Q.J. R. Meteorol. Soc.* **2013**, *139*, 1731–1746.

41. Yano, J.-I.; Plant, R.S. Finite departure from convective quasi-equilibrium: Periodic cycle and discharge-recharge mechanism. *Quator. J. Roy. Meteor. Soc* **2012**, doi:10.1002/qj.957.

42. Plant, R.S.; Yano, J.-I. The energy-cycle analysis of the interactions between shallow and deep atmospheric convection. *Dyn. Atmos. Ocean* **2013**, *64*, 27–52.

43. Yano, J.-I.; Cheedela, S.K.; Roff, G.L. Towards compressed super–parameterization: Test of NAM–SCA under single–column GCM configurations. *Atmos. Chem. Phys. Discuss.* **2012**, *12*, 28237–28303.

44. Randall, D.A.; Pan, D.-M. Implementation of the Arakawa-Schubert cumulus parameterization with a prognostic closure. In *The Representation of Cumulus Convection in Numerical Models*; Emanuel, K.A., Raymond, D.J., Eds.; American Meteor Society: Boston, MA, USA, 1993; pp. 137–144.

45. Pan, D.-M.; Randall, D.A. A cumulus parameterization with prognostic closure. *Quart. J. R. Meteor. Soc.* **1998**, *124*, 949–981.

46. Chen, D.H.; Bougeault, P. A simple prognostic closure assumption to deep convective parameterization. *Acta Meteorol. Sin.* **1992**, *7*, 1–18.

47. Wagner, T.M.; Graf, H.F. An ensemble cumulus convection parameterisation with explicit cloud treatment. *J. Atmos. Sci.* **2010**, *67*, 3854–3869.

48. Plant, R.S.; Yano, J.-I. Comment on "An ensemble cumulus convection parameterisation with explicit cloud treatment" by T.M. Wagner and H.-F. Graf. *J. Atmos. Sci.* **2011**, *68*, 1541–1544.

49. Kuo, H.L. Further studies of the parameterization of the influence of cumulus convection on the large–scale flow. *J. Atmos. Sci.* **1974**, *31*, 1232–1240.

50. Yano, J.-I., and R. S. Plant, 2014: Closure. In *Parameterization of Atmospheric Convection*; Plant, R.S., Yano, J.I., Eds.; Imperial College Press: London, UK, 2014; Volume I, in press.

51. Plant, R.S.; Bengtsson, L. Stochastic aspects of convection parameterization. In *Parameterization of Atmospheric Convection*; Plant, R.S., Yano, J.I., Eds.; Imperial College Press: London, UK, 2014; Volume II, in press.

52. Grant, A.L.M. Cloud–base fluxes in the cumulus–capped boundary layer. *Quart. J. R. Meteor. Soc.* **2001**, *127*, 407–421.

53. Neggers, R.A.J.; Siebesma, A.P.; Lenderink, G.; Holtslag, A.A.M. An evaluation of mass flux closures for diurnal cycles of shallow cumulus. *Mon. Wea. Rev.* **2004**, *132*, 2525–2538.

54. Soares, P.M.M.; Miranda, P.M.A.; Siebesma, A.P.; Teixeira, J. An eddy-diffusivity/mass-flux parametrization for dry and shallow cumulus convection. *Quart. J. R. Meteor. Soc.* **2004**, *130*, 3365–3383.

55. Penland, C. A stochastic approach to nonlinear dynamics: A review. *Bull. Am. Meteor. Soc.* **2003**, *84*, 925–925.

56. Melbourne, I.; Stuart, A. A note on diffusion limits of chaotic skew product flows. *Nonlinearity* **2011**, *24*, 1361–1367.

57. Gottwald, G.A.; Melbourne, I. Homogenization for deterministic maps and multiplicative noise. *Proc. R. Soc. A* **2013**, doi:10.1098/rspa.2013.0201.

58. De Rooy, W.C.; Bechtold, P.; Fröhlich, K.; Hohenegger, C.; Jonker, H.; Mironov, D.; Siebesma, A.P.; Teixeira, J.; Yano, J.-I. Entrainment and detrainment in cumulus convection: An overview. *Quart. J. R. Meteor. Soc.* **2013**, *139*, 1–19.

59. Romps, D.M. A direct measurement of entrainment. *J. Atmos. Sci.* **2010**, *67*, 1908–1927.

60. Dawe, J.T.; Austin, P.H. The influence of the cloud shell on tracer budget measurements of LES cloud entrainment. *J. Atmos. Sci.* **2011**, *68*, 2909–2920.

61. Siebesma, A.P.; Cuijpers, J.W.M. Evaluation of parametric assumptions for shallow cumulus convection. *J. Atmos. Sci.* **1995**, *52*, 650–666.

62. Swann, H. Evaluation of the mass–flux approach to parameterizing deep convection. *Quart. J. R. Meteor. Soc.* **2001**, *127*, 1239–1260.

63. Yano, J.-I.; Guichard, F.; Lafore, J.-P.; Redelsperger, J.-L.; Bechtold, P. Estimations of massfluxes for cumulus parameterizations from high-resolution spatial data. *J. Atmos. Sci.* **2004**, *61*, 829–842.

64. Yano, J.-I.; Baizig, H. Single SCA-plume dynamics. *Dyn. Atmos. Ocean.* **2012**, *58*, 62–94.

65. Korczyka, P.M.; Kowalewskia, T.A.; Malinowski, S.P. Turbulent mixing of clouds with the environment: Small scale two phase evaporating flow investigated in a laboratory by particle image velocimetry. *Physica D* **2011**, *241*, 288–296.

66. Diwan, S.S.; Prasanth, P.; Sreenivas, K.R.; Deshpande, S.M.; Narasimha, R. Cumulus-type flows in the laboratory and on the computer: Simulating cloud form, evolution and large-scale structure. *Bull. Am. Meteor. Soc.* **2014**, doi:10.1175/BAMS-D-12-00105.1.

67. Squires, P. The spatial variation of liquid water content and droplet concentration in cumuli. *Tellus* **1958**, *10*, 372–380.

68. Squires, P. Penetrative downdraughts in cumuli. *Tellus* **1958**, *10*, 381–389.

69. Paluch, I.R. The entrainment mechanism of Colorado cumuli. *J. Atmos. Sci.* **1979**, *36*, 2467–2478.

70. Kain, J.S.; Fritsch, J.L. A one-dimensional entraining/detraining plume model and its application in convective parameterization. *J. Atmos. Sci.* **1990**, *47*, 2784–2802.

71. Bechtold, P.; Köhler, M.; Jung, T.; Doblas-Reyes, F.; Leutbecher, M.; Rodwell, M.; Vitart, F.; Balsamo, G. Advances in simulating atmospheric variability with the ECMWF model: From synoptic to decadal time-scales. *Quart. J. R. Meteor. Soc.* **2008**, *134*, 1337–1351.

72. Derbyshire, S.H.; Maidens, A.V.; Milton, S.F.; Stratton, R.A.; Willett, M.R. Adaptive detrainment in a convective parameterization. *Quart. J. R. Meteor. Soc.* **2011**, *137*, 1856–1871.

73. De Rooy, W.C.; Siebesma, A.P. A simple parameterization for detrainment in shallow cumulus. *Mon. Wea. Rev.* **2008**, *136*, 560–576.

74. Böing, S.J.; Siebesma, A.P.; Korpershoek, J.D.; Jonker, H.J.J. Detrainment in deep convection. *Geophys. Res. Lett.* **2012**, doi:10.1029/2012GL053735.

75. De Rooy, W.C.; Siebesma, A.P. Analytical expressions for entrainment and detrainment in cumulus convection. *Quart. J. R. Meteor. Soc.* **2010**, *136*, 1216–1227.

76. Neggers, R.A.J. A dual mass flux framework for boundary layer convection. Part II: Clouds. *J. Atmos. Sci.* **2009**, *66*, 1489–1506.

77. Morton, B.R.; Taylor, G.; Turner, J.S. Turbulent gravitational convection from maintained and instantaneous sources. *Proc. Roy. Meteor. Soc.* **1956**, *234*, 1–33.

78. Morton, B.R. Discreet dry convective entities: II Thermals and deflected jets. In *The Physics and Parameterization of Moist Atmospheric Convection*; Smith, R.K., Ed.; NATO ASI, Kloster Seeon, Kluwer Academic Publishers: Dordrecht, The Netherlands, 1997; pp. 175–210.

79. Yano, J.-I. Basic convective element: Bubble or plume?: A historical review. *Atmos. Phys. Chem. Dis.* **2014**, *14*, 3337–3359.

80. Squires, P.; Turner, J.S. An entraining jet model for cumulo–nimbus updraught. *Tellus* **1962**, *16*, 422–434.

81. Bringi, V.N.; Knupp, K.; Detwiler, A.; Liu, L.; Caylor, I.J.; Black, R.A. Evolution of a Florida thunderstorm during the Convection and Precipitation/Electrification Experiment: The case of 9 August 1991. *Mon. Wea. Rev.* **1997**, *125*, 2131–2160.

82. Jonker, H.; His Collaborators. University of Deft, Deft, the Netherlands. Personal communication, 2009.

83. Heus, T.; Jonker, H.J.J.; van den Akker, H.E.A.; Griffith, E.J.; Koutek, M.; Post, F.H. A statistical approach to the life cycle analysis of cumulus clouds selected in a vitual reality environment. *J. Geophys. Res.* **2009**, doi:10.1029/2008JD010917.

84. Levine, J. Spherical vortex theory of bubble-like motion in cumulus clouds. *J. Meteorol.* **1959**, *16*, 653–662.

85. Gerard, L.; Geleyn, J.-F. Evolution of a subgrid deep convection parameterization in a limited–area model with increasing resolution. *Quart. J. R. Meteor. Soc.* **2005**, *131*, 2293–2312.

86. Stanley, H.E. *Introduction to Phase Transitions and Critical Phenomena*; Oxford University Press: London, UK, 1971.

87. Vattay, G.; Harnos, A. Scaling behavior in daily air humidity fluctuations. *Phys. Rev. Lett.* **1994**, *73*, 768–771.

88. Peters, O.; Hertlein, C.; Christensen, K. A complexity view of rainfall. *Phys. Rev. Lett.* **2002**, *88*, 018701.

89. Newman, M.E.J. Power laws, Pareto distributions and Zipf's law. *Cont. Phys.* **2005**, *46*, 323–351.

90. Peters, O.; Neelin, J.D. Critical phenomena in atmospheric precipitation. *Nat. Phys* **2006**, *2*, 393–396.

91. Peters, O.; Deluca, A.; Corral, A.; Neelin, J.D.; Holloway, C.E. Universality of rain event size distributions. *J. Stat. Mech.* **2010**, *11*, P11030.

92. J. D. Neelin, O. Peters, J. W.-B. Lin, K. Hales, and C. E. Holloway. Rethinking convective quasi-equilibrium: Observational constraints for stochastic convective schemes in climate models. *Phil. Trans. R. Soc. A* **2008**, *366*, 2581–2604.

93. Peters, O.; Neelin, J.D.; Nesbitt, S.W. Mesoscale convective systems and critical clusters. *J. Atmos. Sci.* **2009**, *66*, 2913–2924.

94. Wood, R.; Field, P.R. The distribution of cloud horizontal sizes. *J. Clim.* **2011**, *24*, 4800–4816.

95. Muller, C.J.; Back, L.E.; O'Gorman, P.A.; Emanuel, K.A. A model for the relationship between tropical precipitation and column water vapor. *Geophys. Res. Lett.* **2009**, doi:10.1029/2009GL039667.

96. Krueger, S. University of Utah, Salt Lake, UT, USA. Personal communication, 2014.

97. Seo, E.-K.; Sohn, B.-J.; Liu, G. How TRMM precipitation radar and microwave imager retrieved rain rates differ. *Geophys. Res. Lett.* **2007**, doi:10.1029/2007GL032331.

98. Yano, J.-I.; Liu, C.; Moncrieff, M.W. Atmospheric convective organization: Homeostasis or self-organized criticality? *J. Atmos. Sci.* **2012**, *69*, 3449–3462,

99. Raymond, D.J. Thermodynamic control of tropical rainfall. *Quart. J. Roy. Meteor. Soc.* **2000**, *126*, 889–898.

100. Peters, O.; Pruessner, G. Tuning- and Order Parameter in the SOC Ensemble. arXiv:0912.2305v1, 11 December 2009.

101. Yano, J.-I.; Redelsperger, J.-L.; Guichard, F.; Bechtold, P. Mode decomposition as a methodology for developing convective-scale representations in global models. *Quart. J. R. Meteor. Soc.* **2005**, *131*, 2313–2336.

102. Yano, J.-I.; Benard, P.; Couvreux, F.; Lahellec, A. NAM–SCA: Nonhydrostatic Anelastic Model under Segmentally–Constant Approximation. *Mon. Wea. Rev.* **2010**, *138*, 1957–1974.

103. Yano, J.-I. Mass–flux subgrid–scale parameterization in analogy with multi–component flows: A formulation towards scale independence. *Geosci. Model Dev.* **2012**, *5*, 1425–2440.

104. Arakawa, A.; Wu, C.-M. A unified representation of deep moist convection in numerical modeling of the atmosphere. Part I. *J. Atmos. Sci.* **2013**, *70*, 1977–1992.

105. Arakawa, A.; Jung, J.-H.; Wu, C.-M. Toward unification of the multiscale modeling of the atmosphere. *Atmos. Chem. Phys.* **2011**, *11*, 3731–3742.

106. Bihlo, A. A Tutorial on Hamiltonian Mechanics. COST Document, 2011. Available online: http://convection.zmaw.de/fileadmin/user_upload/convection/Convection/COST_Documents/Search_for_New_Frameworks/A_Tutorial_on_Hamiltonian_Mechanics.pdf (accessed on 4 December 2014).

107. Cardoso-Bihlo, E.; Popovych, R.O. Invariant Parameterization Schemes. COST Document, 2012. Available online: http://convection.zmaw.de/fileadmin/user_upload/convection/Convection/COST_Documents/Search_for_New_Frameworks/Invariant_Parameterization_Schemes.pdf (accessed on 4 December 2014).

108. Popovych, R.O.; Bihlo, A. Symmetry perserving parameterization schemes. *J. Math. Phycs.* **2012**, *53*, 073102.

109. Bihlo, A.; Bluman, G. Conservative parameterization schemes. *J. Math. Phys.* **2013**, *54*, 083101.

110. Bihlo, A.; Dos Santos Cardoso-Bihlo, E.M.; Popovych, R.O. Invariant parameterization and turbulence modeling on the beta-plane. *Physica D* **2014**, *269*, 48–62.

111. Grant, A.L.M. Cumulus convection as a turbulent flow. In *Parameterization of Atmospheric Convection*; Plant, R.S.; Yano, J.I., Eds.; Imperial College Press: London, UK, 2014; Volume II, in press.

112. Grant, A.L.M. Precipitating convection in cold air: Virtual potential temperature structure. *Quart. J. R. Meteor. Soc.* **2007**, *133*, 25–36.

113. Seifert, A.; Stevens, B. Microphysical scaling relations in a kinematic model of isolated shallow cumulus clouds. *J. Atmos. Sci.* **2010**, *67*, 1575–1590.

114. Seifert, A.; Zänger, G. Scaling relations in warm-rain orographic precipitation. *Meteorol. Z.* **2010**, *19*, 417–426.

115. Stevens, B.; Seifert, A. Understanding microphysical outcomes of microphysical choices in simulations of shallow cumulus convection. *J. Met. Soc. Jpn.* **2008**, *86A*, 143–162.

116. Yano, J.-I.; Phillips, V.T.J. Ice–ice collisions: An ice multiplication process in atmospheric clouds. *J. Atmos. Sci.* **2011**, *68*, 322–333.

117. Chandrasekahr, S. *Hydrodynamic and Hydromagnetic Instability*; Oxford University Press: London, UK, 1961.

118. Marquet, P.; Geleyn, J.-F. On a general definition of the squared Brunt–Väisälä frequency associated with the specific moist entropy potential temperature. *Quart. J. R. Meteor. Soc.* **2013**, *139*, 85–100.

119. Durran, D.R.; Klemp, J.B. On the effects of moisture on the Brunt-Vaisala frequency. *J. Atmos. Sci.* **1982**, *39*, 2152–2158.

120. Machulskaya, E. Clouds and convection as subgrid–scale distributions. In *Parameterization of Atmospheric Convection*; Plant, R.S., Yano, J.I., Eds.; Imperial College Press: London, UK, 2014; Volume II, in press.

121. Bony, S.; Emanuel, K.A. A parameterization of the cloudness associated with cumulus convection: Evaluation using TOGA COARE data. *J. Atmos. Sci.* **2001**, *58*, 3158–3183.

122. Tompkins, A.M. A prognostic parameterization for the subgrid–scale variability of water vapor and clouds in large–scale models and its use to diagnose cloud cover. *J. Atmos. Sci.* **2002**, *59*, 1917–1942.

123. Larson, V.E.; Wood, R.; Field, P.R.; Goraz, J.-C.; Haar, T.H.V.; Cotton, W.R. Systematic biases in the microphysics and thermodynamics of numerical models that ignore subgrid–scale variability. *J. Atmos. Sci.* **2002**, *58*, 1117–1128.

124. Wilson, D.R.; Bushell, A.C.; Kerr-Munslow, A.M.; Price, J.D.; Morcrette, C.J. PC2: A prognostic cloud fraction and condensation scheme. I: Scheme description. *Quart. J. R. Meteor. Soc.* **2008**, *134*, 2093–2107.

125. Klein, S.A.; Pincus, R.; Hannay, C.; Xu, K.-M. How might a statistical cloud scheme be coupled to a mass–flux convection scheme? *J. Geophys. Res.* **2005**, *110*, D15S06.

126. Golaz, J.-C.; Larson, V.E.; Cotton, W.R. A PDF–based model for boundary layer clouds. Part I: Method and model description. *J. Atmos. Sci.* **2002**, *59*, 3540–3551.

127. Plant, R. S. Statistical properties of cloud lifecycles in cloud-resolving models. *Atmos. Chem. Phys.* **2009**, *9*, 2195–2205.

128. Sakradzija, M.; Seifert, A.; Heus, T. Fluctuations in a quasi-stationary shallow cumulus cloud ensemble, *Nonlin. Processes Geophys. Discuss.* **2014**, *1*, 1223–1282.

129. Marquet, P. On the definition of a moist-air potential vorticity. *Quart. J. R. Meteor. Soc.* **2014**, *140*, 917–929.

130. Gerard, L. Model resolution issues and new approaches in the grey zone. In *Parameterization of Atmospheric Convection*; Plant, R.S., Yano, J.I., Eds.; Imperial College Press: London, UK, 2014; Volume II, in press.

131. Bechtold, P. ECMWF, Reading, UK. Personal communication, 2014.

132. Bengtsson, L.; Körnich, H.; Källén, E.; Svensson, G. Large-scale dynamical response to subgrid scale organization provided by cellular automata. *J. Atmos. Sci.* **2011**, *68*, 3132–3144.

133. Bengtsson, L.; Steinheimer, M.; Bechtold, P.; Geleyn, J.-F. A stochastic parameterization using cellular automata. *Quart. J. R. Meteor. Soc.* **2013**, *139*, 1533–1543.

134. Gerard, L. An integrated package for subgrid convection, clouds and precipitation compatible with the meso-gamma scales. *Quart. J. R. Meteor. Soc.* **2007**, *133*, 711–730

135. Gerard, L.; Piriou, J.-M.; Brožková, R.; Geleyn, J.-F.; Banciu, D. Cloud and precipitation parameterization in a meso-gamma-scale operational weather prediction model. *Mon. Weather Rev.* **2009**, *137*, 3960–3977.

136. Plant, R.S.; Craig, G.C. A stochastic parameterization for deep convection based on equilibrium statistics. *J. Atmos. Sci.* **2008**, *65*, 87–105.

137. Machulskaya, E.; Mironov, D. Implementation of TKE–Scalar Variance Mixing Scheme into COSMO. Available online: http://www.cosmo-model.org (accessed on 4 December 2014).

138. Mironov, D. Turbulence in the lower troposphere: Second-order closure and mass-flux modelling frameworks. *Lect. Notes. Phys.* **2009**, *756*, 161–221.

139. Smagorinsky, J. General circulation experiments with the primitive equations. I. The basic equations. *Mon. Wea. Rev.* **1963**, *91*, 99–164.

140. Jaynes, E.T. *Probability Theory, The Logic of Science*; Cambridge University Press: Cambridge, UK, 2003.

141. Gregory, P.C. *Bayesian Logical Data Analysis for the Physical Sciences*; Cambridge University Press: Cambridge, UK, 2005.

142. Khain, A.P.; Beheng, K.D.; Heymsfield, A.; Korolev, A.; Krichak, S.O.; Levin, Z.; Pinsky, M.; Phillips, V.; Prabhakaran, T.; Teller, A.; *et al.* Representation of microphysical processes in cloud–resolving models: Spectral (bin) microphysics *vs.* bulk–microphysics. *Rev. Geophys.* **2014**, submitted.

143. Phillips, V.T.J. Microphysics of convective cloud and its treatment in parameterization. In *Parameterization of Atmospheric Convection*; Plant, R.S., Yano, J.I., Eds.; Imperial College Press: London, UK, 2014; Volume II, in press.

144. Khain, A.P.; Lynn, B.; Shpund, J. Simulation of intensity and structure of hurricane Irene: Effects of aerosols, ocean coupling and model resolution. In Proceedings of the 94-th AMS Conference, Atlanta, GA, USA, 2–6 February 2014.

145. Lynn, B.H.; Khain, A.P.; Bao, J.W.; Michelson, S.A.; Yuan, T.; Kelman, G.; Shpund, J.; Benmoshe, N. The Sensitivity of the Hurricane Irene to aerosols and ocean coupling: Simulations with WRF with spectral bin microphysics. *J. Atmos. Sci.* **2014**, submitted.

146. Phillips, V.T.J.; Khain, A.; Benmoshe, N.; Ilotovich, E. Theory of time-dependent freezing and its application in a cloud model with spectral bin microphysics. I: Wet growth of hail. *J. Atmos. Sci.* **2014**, in press.

147. Phillips, V.T.J.; Khain, A.; Benmoshe, N.; Ilotovich, E.; Ryzhkov, A. Theory of time-dependent freezing and its application in a cloud model with spectral bin microphysics. II: Freezing raindrops and simulations. *J. Atmos. Sci.* **2014**, in press.

148. Ilotoviz, E.; Khain, A.P.; Phillips, V.T.J.; Benmoshe, N.; Ryzhkov, A.V. Effect of aerosols on formation and growth regime of hail and freezing drops. *J. Atmos. Sci.* **2014**, submitted.

149. Kumjian, M.R.; Khain, A.P.; Benmoshe, N.; Ilotoviz, E.; Ryzhkov, A.V.; Phillips, V.T.J. The anatomy and physics of ZDR columns: Investigating a polarimetric radar signature with a spectral bin microphysical model. *J. Appl. Meteorol. Climatol.* **2014**, in press.

150. Khain, A.P.; Ilotoviz, E.; Benmoshe, N.; Phillips, V.T.J.; Kumjian, M.R.; Ryzhkov, A.V. 2014 Application of the theory of time-dependent freezing and hail growth to analysis of hail storm using a cloud model with SBM. In Proceedings of the 94-th AMS Conference, Atlanta, GA, USA, 2–6 February 2014.

151. Zipser, E.J. The role of unsaturated convective downdrafts in the structure and rapid decay of an equatorial disturbance. *J. Appl. Met.* **1969**, *8*, 799–814.

152. Zipser, E.J. Mesoscale and convective–scale downdrafts as distinct components of squall–line circulation. *Mon. Wea. Rev.* **1977**, *105*, 1568–1589.

153. Houze, R.A., Jr.; Betts, A.K. Convection in GATE. *Rev. Geophys. Space Phys.* **1981**, *19*, 541–576.

154. Emanuel, K.A. The finite-amplitude nature of tropical cyclogenesis. *J. Atmos. Sci.* **1989**, *46*, 3431–3456.

155. Yano, J.-I.; Emanuel, K.A. An improved model of the equatorial troposphere and its coupling with the stratosphere. *J. Atmos. Sci.* **1991**, *48*, 377–389.

156. Fritsch, J.M.; Chappell, C.F. Numerical prediction of convectively driven mesoscale pressure systems: Part I: Convective parameterization. *J. Atmos. Sci.* **1980**, *37*, 1722–1733.

157. Tiedtke, M. A comprehensive mass flux scheme of cumulus parameterization in large–scale models. *Mon. Wea. Rev.* **1989**, *117*, 1779–1800.

158. Zhang, G.J.; McFarlane, N.A. Sensitivity of climate simulations of the parameterization of cumulus convection in the Canadian Climate Centre general circulation model. *Atmos. Ocean* **1995**, *33*, 407–446.

159. Bechtold, P.; Bazile, E.; Guichard, F.; Mascart, P.; Richard, E. A mass-flux convection scheme for regional and global models. *Quart. J. R. Meteor. Soc.* **2001**, *127*, 869–889.

160. Yano, J.-I. Downdraught. In *Parameterization of Atmospheric Convection*; Plant, R.S., Yano, J.I., Eds.; Imperial College Press: London, UK, 2014; Volume I; in press.

161. Warner, C.; Simpson, J.; Martin, D.W.; Suchman, D.; Mosher, F.R.; Reinking, R.F. Shallow convection on day 261 of GATE: Mesoscale arcs. *Mon. Wea. Rev.* **1979**, *107*, 1617–1635.

162. Zuidema, P.; Li, Z.; Hill, R.J.; Bariteau, L.; Rilling, B.; Fairall, C. Brewer, W.A.; Albrecht, B.; Hare, J. On trade wind cumulus cold pools. *J. Atmos. Sci.* **2012**, *69*, 258–280.

163. Barnes, G.M.; Garstang, M. Subcloud layer energetics of precipitating convection. *Mon. Wea. Rev.* **1982**, *110*, 102–117.

164. Addis, R.P.; Garstang, M.; Emmitt, G.D. Downdrafts from tropical oceanic cumuli. *Bound. Layer Meteorol.* **1984**, *28*, 23–49.

165. Young, G.S.; Perugini, S.M.; Fairall, C.W. Convective wakes in the Equatorial Western Pacific during TOGA. *Mon. Wea. Rev.* **1995**, *123*, 110–123.

166. Lima, M.A.; Wilson, J.W. Convective storm initiation in a moist tropical environment. *Mon. Wea. Rev.* **2008**, *136*, 1847–1864.

167. Flamant, C.; Knippertz, P.; Parker, D.J.; Chaboureau, J.-P.; Lavaysse, C.; Agusti-Panareda, A.; Kergoat, L. The impact of a mesoscale convective system cold pool on the northward propagation of the intertropical discontinuity over West Africa. *Quart. J. R. Meteor. Soc.* **2009**, *135*, 139–159.

168. Lothon, M.; Campistron, B.; Chong, M.; Couvreux, F.; Guichard, F.; Rio, C.; Williams, E. Life cycle of a mesoscale circular gust front observed by a C-band doppler radar in West Africa. *Mon. Wea. Rev.* **2011**, *139*, 1370–1388.

169. Dione, C.; Lothon, M.; Badiane, D.; Campistron, B.; Couvreux, F.; Guichard, F.; Sall, S.M. Phenomenology of Sahelian convection observed in Niamey during the early monsoon. *Quart. J. R. Meteor. Soc.* **2013**, doi:10.1002/qj.2149.

170. Khairoutdinov, M.; Randall, D. High-resolution simulations of shallow-to-deep convection transition over land. *J. Atmos. Sci.* **2006**, *63*, 3421–3436.

171. Böing, S.J.; Jonker, H.J.J.; Siebesma, A.P.; Grabowski, W.W. Influence of the subcloud layer on the development of a deep convective ensemble. *J. Atmos. Sci.* **2012**, *69*, 2682–2698.

172. Schlemmer, L.; Hohenegger, C. The formation of wider and deeper clouds as a result of cold-pool dynamics. *J. Atmos. Sci.* **2014**, submitted.

173. Grandpeix, J-Y.; Lafore, J.-P. A density current parameterization coupled with emanuel's convection scheme. Part I: The models. *J. Atmos. Sci.* **2010**, *67*, 881–897.

174. Yano, J.-I. Comments on "A density current parameterization coupled with emanuel's convection scheme. Part I: The models". *J. Atmos. Sci.* **2012**, *69*, 2083–2089.

175. Browning, K.A.; Hill, F.F.; Pardoe, C.W. Structure and mechanism of precipitation and the effect of orography in a wintertime warm sector. *Quart. J. Roy. Meteor. Soc.* **1974**, *100*, 309–330.

176. Fuhrer, O.; Schär, C. Embedded cellular convection in moist flow past topography. *J. Atmos. Sci.* **2005**, *62*, 2810–2828.

177. Yano, J.-I. Downscaling, parameterization, decomposition, compression: A perspective from the multiresolutional analysis. *Adv. Geophy.* **2010**, *23*, 65–71.

178. Rezacova, D.; Szintai, B.; Jakubiak, B.; Yano, J.I.; Turner, S. Verification of high resolution precipitation forecast by radar-based data. In *Parameterization of Atmospheric Convection*; Plant, R.S., Yano, J.I., Eds.; Imperial College Press: London, UK, 2014; Volume II, in press.

179. Vukicevic, T.; Rosselt, D. Analysis of the impact of model nonlinearities in inverse problem solving. *J. Atmos. Sci.* **2008**, *65*, 2803–2823.

180. Rosselt, D.; Bishop, C.H. Nonlinear parameter estimation: Comparison of an ensemble Kalman smoother with a Markov chain Monte Carlo algorithm. *Mon. Wea. Rev.* **2012**, *140*, 1957–1974.

181. Khain, A. Notes on state-of-the-art investigations of aerosol effects on precipitation: A critical review. *Environ. Res. Lett.* **2008**, doi:10.1088/1748-9326/1/015004.

182. Rosenfeld, D.; Lohmann, U.; Raga, G.B.; O'Dowd, C.D.; Kulmala, M.; Fuzzi, S.; Reissell, A.; Andreae, M.O. Flood or drought: How do aerosols affect precipitation. *Science* **2008**, *321*, 1309–1313.

183. Boucher, O.; Quaas, J. Water vapour affects both rain and aerosol optical depth. *Nat. Geosci.* **2013**, *6*, 4–5.

184. Freud, E.; Rosenfeld, D. Linear relation between convective cloud drop number concentration and depth for rain initiation. *J. Geophys. Res.* **2012**, *117*, D02207.

185. Mewes, D. Test Einer Neuen Parametrisierung Konvektiven Niederschlags Im Klimamodell. Bachelor's Thesis, University of Leipzig, Leipzig, Germany, 2013.

186. Lohmann, U.; Quaas, J.; Kinne, S.; Feichter, J. Different approaches for constraining global climate models of the anthropogenic indirect aerosol effect. *Bull. Am. Meteorol. Soc.* **2007**, *88*, 243–249.

187. Bodas-Salcedo, A.; Webb, M.J.; Bony, S.; Chepfer, H.; Dufresne, J.-L.; Klein, S.A.; Zhang, Y.; Marchand, R.; Haynes, J.M.; Pincus, R.; *et al.* COSP: Satellite simulation software for model assessment. *Bull. Amer. Meteor. Soc.* **2011**, *92*, 1023–1043.

188. Nam, C.; Quaas, J. Evaluation of clouds and precipitation in the ECHAM5 general circulation model using CALIPSO and CloudSat. *J. Clim.* **2012**, *25*, 4975–4992.

189. Nam, C.; Quaas, J.; Neggers, R.; Drian, C.S.; Isotta, F. Evaluation of boundary layer cloud parameterizations in the ECHAM5 general circulation model using CALIPSO and CloudSat satellite data. *J. Adv. Model. Earth Syst.* **2014**, doi:10.1002/2013MS000277.

190. Gehlot, S.; Quaas, J. Convection-climate feedbacks in ECHAM5 general circulation model: A Lagrangian trajectory perspective of cirrus cloud life cycle. *J. Clim.* **2012**, *25*, 5241–5259.

191. Suzuki, K.; Stephens, G.L.; van den Heever, S.C.; Nakajima, T.Y. Diagnosis of the warm rain process in cloud-resolving models using joint CloudSat and MODIS observations. *J. Atmos. Sci.* **2011**, *68*, 2655–2670.

192. Schirber, S.; Klocke, D.; Pincus, R.; Quaas, J.; Anderson, J. Parameter estimation using data assimilation in an atmospheric general circulation model: From a perfect towards the real world, *J. Adv. Model. Earth Syst.* **2013**, *5*, 1942–2466.

193. Yano, J.-I.; Lane, T.P. Convectively-generated gravity waves simulated by NAM-SCA. *J. Geophys. Res.* **2014**, doi:10.1002/2013JD021419.

194. Tobias, S.M.; Marston, J.B. Direct Statistical simulation of out-of-equilibrium jets. *Phys. Rev. Lett.* **2013**, *110*, 104502.

Permissions

All chapters in this book were first published in Atmosphere, by MDPI; hereby published with permission under the Creative Commons Attribution License or equivalent. Every chapter published in this book has been scrutinized by our experts. Their significance has been extensively debated. The topics covered herein carry significant findings which will fuel the growth of the discipline. They may even be implemented as practical applications or may be referred to as a beginning point for another development.

The contributors of this book come from diverse backgrounds, making this book a truly international effort. This book will bring forth new frontiers with its revolutionizing research information and detailed analysis of the nascent developments around the world.

We would like to thank all the contributing authors for lending their expertise to make the book truly unique. They have played a crucial role in the development of this book. Without their invaluable contributions this book wouldn't have been possible. They have made vital efforts to compile up to date information on the varied aspects of this subject to make this book a valuable addition to the collection of many professionals and students.

This book was conceptualized with the vision of imparting up-to-date information and advanced data in this field. To ensure the same, a matchless editorial board was set up. Every individual on the board went through rigorous rounds of assessment to prove their worth. After which they invested a large part of their time researching and compiling the most relevant data for our readers.

The editorial board has been involved in producing this book since its inception. They have spent rigorous hours researching and exploring the diverse topics which have resulted in the successful publishing of this book. They have passed on their knowledge of decades through this book. To expedite this challenging task, the publisher supported the team at every step. A small team of assistant editors was also appointed to further simplify the editing procedure and attain best results for the readers.

Apart from the editorial board, the designing team has also invested a significant amount of their time in understanding the subject and creating the most relevant covers. They scrutinized every image to scout for the most suitable representation of the subject and create an appropriate cover for the book.

The publishing team has been an ardent support to the editorial, designing and production team. Their endless efforts to recruit the best for this project, has resulted in the accomplishment of this book. They are a veteran in the field of academics and their pool of knowledge is as vast as their experience in printing. Their expertise and guidance has proved useful at every step. Their uncompromising quality standards have made this book an exceptional effort. Their encouragement from time to time has been an inspiration for everyone.

The publisher and the editorial board hope that this book will prove to be a valuable piece of knowledge for researchers, students, practitioners and scholars across the globe.

List of Contributors

Barbara Kozielska
Department of Air Protection, Silesian University of Technology, 2 Akademicka St., 44-100 Gliwice, Poland

Wioletta Rogula-Kozłowska and Krzysztof Klejnowski
Institute of Environmental Engineering, Polish Academy of Sciences, 34 M. Skłodowskiej-Curie St., 41-819 Zabrze, Poland

Marco Gabella, Maurizio Sartori, Marco Boscacci and Urs Germann
Meteo Swiss, via ai Monti 146, Locarno Monti CH-6605, Switzerland

Yan Fang and Jie Yin
School of Tourism and City Management, Zhejiang Gongshang University, 18 Xuezheng Street, Hangzhou 310018

Yang Li
Center for Atmosphere Watch and Service, Meteorological Observation Center, China Meteorological Administration, Beijing 100081, China

Quanliang Chen, Lin Wang and Ran Tao
Plateau Atmospheric and Environment Key Laboratory of Sichuan Province, College of Atmospheric Sciences, Chengdu University of Information Technology, Chengdu 610225, China

Hujia Zhao
Institute of Atmospheric Environment, China Meteorological Administration, Shenyang 110016, China

Qiyuan Wang, Suixin Liu, Yaqing Zhou, Yongming Han, Haiyan Ni and Ningning Zhang
Key Laboratory of Aerosol Chemistry & Physics, Institute of Earth Environment, Chinese Academy of Sciences, Xi'an 710061, China

Junji Cao
Key Laboratory of Aerosol Chemistry & Physics, Institute of Earth Environment, Chinese Academy of Sciences, Xi'an 710061, China
Institute of Global Environmental Change, Xi'an Jiaotong University, Xi'an 710049, China

Rujin Huang
Key Laboratory of Aerosol Chemistry & Physics, Institute of Earth Environment, Chinese Academy of Sciences, Xi'an 710061, China
Lab of Atmospheric Chemistry, Paul Scherrer Institute (PSI), Villigen 5232, Switzland
Centre for Climate and Air Pollution Studies, Ryan Institute, National University of Ireland Galway, University Road, Galway, Ireland

Xingchang Wang and Chuankuan Wang
Center for Ecological Research, Northeast Forestry University, 26 Hexing Road, Harbin 150040, China

Qinglin Li
School of Forestry and Bio-technology, Zhejiang A & F University, 88 North Road of Huancheng, Lin'an 311300, China
Forest Analysis and Inventory Branch, Ministry of Forests, Lands, and Natural Resource Operations, 727 Fisgard Street, Victoria, BC V8W-9C2, Canada

Huanbo Wang, Guangming Shi and Fumo Yang
Key Laboratory of Reservoir Aquatic Environment of CAS, Chongqing Institute of Green and Intelligent Technology, Chinese Academy of Sciences, Chongqing 400714, China

Xinghua Li
School of Chemistry and Environment, Beihang University, Beijing 100191, China

Junji Cao
Key Lab of Aerosol Chemistry & Physics, Institute of Earth Environment, Chinese Academy of Sciences, Xi'an 710075, China

Chengcai Li
School of Physics, Peking University, Beijing 100871, China

Yongliang Ma and Kebin He
State Environmental Protection Key Laboratory of Sources and Control of Air Pollution Complex, School of Environment, Tsinghua University, Beijing 100084, China

Sonia Wharton, Matthew Simpson and Jessica L. Osuna
Atmospheric, Earth and Energy Division, Lawrence Livermore National Laboratory, 7000 East Avenue, L-103, Livermore, CA 94550, USA

Jennifer F. Newman
School of Meteorology, University of Oklahoma, 120 David L. Boren Blvd., Norman, OK 73072, USA

Sebastien C. Biraud
Earth Sciences Division, Lawrence Berkeley National Laboratory, 1 Cyclotron Rd., Berkeley, CA 94720, USA

Jun-Ichi Yano, Jean-François Geleyn and Pascal Marquet
CNRM/GAME UMR 3589, Météo-France and CNRS, 31057 Toulouse Cedex, France

Martin Köhler and Dmitrii Mironov
DWD, 63067 Offenbach, Germany

Johannes Quaas
Institute for Meteorology, Universität Leipzig, 04103 Leipzig, Germany

Pedro M. M. Soares
Instituto Dom Luiz, Faculdade de Ciências, Universidade de Lisboa, 1749-016 Lisboa, Portugal

Vaughan T. J. Phillips
Department of Physical Geography,and Ecosystem Science, University of Lund, Solvegatan 12, Lund 223 62, Sweden

Robert S. Plant
Department of Meteorology, University of Reading, RG6 6BB Reading, UK;

Anna Deluca
Max Planck Institute for the Physics of Complex Systems, 01187 Dresden, Germany

Lukrecia Stulic and Zeljka Fuchs
Department of Physics, University of Split, 21000 Split, Croatia

Printed in the USA
CPSIA information can be obtained
at www.ICGtesting.com
JSHW051438221024
72173JS00006B/1506